Statistical Physics of
DNA

An Introduction to Melting, Unzipping and Flexibility of the Double Helix

Statistical Physics of
DNA

An Introduction to Melting, Unzipping and Flexibility of the Double Helix

Nikos Theodorakopoulos

National Hellenic Research Foundation, Greece &
University of Konstanz, Germany

World Scientific

NEW JERSEY · LONDON · SINGAPORE · BEIJING · SHANGHAI · HONG KONG · TAIPEI · CHENNAI · TOKYO

Published by

World Scientific Publishing Co. Pte. Ltd.

5 Toh Tuck Link, Singapore 596224

USA office: 27 Warren Street, Suite 401-402, Hackensack, NJ 07601

UK office: 57 Shelton Street, Covent Garden, London WC2H 9HE

Library of Congress Control Number: 2019057638

British Library Cataloguing-in-Publication Data
A catalogue record for this book is available from the British Library.

STATISTICAL PHYSICS OF DNA
An Introduction to Melting, Unzipping and Flexibility of the Double Helix

ISBN 978-981-120-953-6

For any available supplementary material, please visit
https://www.worldscientific.com/worldscibooks/10.1142/11533#t=suppl

To Tessa

Foreword

The DNA molecule is associated with an unusually broad range of cooperative physical phenomena, observable at a macro- as well as a mesoscopic scale. With the hindsight of almost 70 years of intensive, global research effort, it is therefore not surprising that many of its properties have proved to be amenable to a theoretical treatment based on statistical physics.

The theory of phase transitions and critical phenomena, developed in the 1960's and 70's transformed our understanding of the variability of the structure of matter. One of the main insights gained was that only a few degrees of freedom and/or interactions between them are responsible for driving a phase transformation. The related concept of a "soft phonon mode" was particularly fruitful in the development of the theory of structural phase transitions.

The stability of the double helix is contingent on fine-tuning a number of physicochemical control parameters. Its unwinding, occurring as a result of varying any one of them, provides an almost unique physical example of a phase transition in a one-dimensional system. As such it was almost predestined to serve as a laboratory for new ideas, which followed — and occasionally paralleled — the development of the general theory of phase transitions. The present book is an attempt to present some of these ideas.

The content of this book evolved from a set of lectures on biological physics given at the University of Konstanz and highlighting developments in single-molecule DNA experiments (e.g. stretching, overstretching, unzipping, hairpin equilibria), and their interpretation in terms of statistical mechanics, to include some further, more specialized material which provides detailed support to the dynamical theory of DNA melting, i.e. the Peyrard–Bishop–Dauxois model. I have intentionally included a considerable measure of computational details, as well as fits to experimental data, because I think that the model's status is far beyond "proof of principle", and can actually be used for a quantitative description of melting in heterogeneous DNA.

My intention was to give a self-contained account of the topics presented. This means that some sections have been expanded to include an appropriate general introduction. I therefore hope that the book can be read by students with an elementary background in statistical mechanics, while still containing enough interesting material for the more advanced reader.

Although I have in most cases attempted to provide credit to the original sources in the literature, I am aware that the scope of the book renders a full success of such an undertaking almost impossible. I must therefore apologize for any serious omissions.

I would like to thank Josef Jäckle (Konstanz) and Michel Peyrard (Lyon) for getting me interested in biophysics and for fruitful discussions over many years.

Nikos Theodorakopoulos

Konstanz & Athens,
October 2018

About the Author

Nikos Theodorakopoulos, born (1947) in Athens, Greece, went to the U.S. on an undergraduate Fulbright scholarship. He attended Harvey Mudd College, the University of Illinois (Urbana) and Brown University, where he completed his Ph.D. in statistical physics under Leo Kadanoff in 1971. He subsequently worked in Germany, first at what is now the Forschungszentrum Jülich, and later at the University of Konstanz, where he obtained his Habilitation degree, and at the Max-Planck Institute for Solid State Research in Stuttgart. He returned to Athens in 1989 to join the Theoretical & Physical Chemistry Institute of the National Hellenic Research Foundation, where he is currently an emeritus Director of Research. Since 2003 he has also been an adjunct professor of theoretical physics at the University of Konstanz.

His research interests cover a broad spectrum of theoretical condensed matter physics (critical phenomena, low-temperature properties of amorphous solids, soliton physics and statistical mechanics of biopolymers).

He has held visiting professorships in Switzerland (University of Basel), Japan (University of Tokyo) and France (ENS de Lyon).

Contents

Foreword   vii

1. Statistical mechanics of simple polymer chain models   1

    1.1   Definitions and notation. Key quantities   1
          1.1.1   The freely jointed chain (FJC)   2
          1.1.2   Local stiffness. Elastic energy. The wormlike chain   2
    1.2   Statistical mechanics of the Kratky–Porod (KP) chain . .   3
          1.2.1   The partition function   3
          1.2.2   Correlations. Persistence length   4
          1.2.3   Moments   5
          1.2.4   Radius of gyration   5
    1.3   The end-to-end distance probability distribution   7
    1.4   The Gaussian chain   11
    1.5   Excluded volume effects   14
          1.5.1   Characterization of the end-to-end distance distri-
                  bution function   14
          1.5.2   Critical exponents and observable quantities . . .   15
          1.5.3   Values of the critical exponents. Relationship to
                  the self-avoiding walk problem   15

2. Entropic elasticity: the DNA force-extension relationship   19

    2.1   Statistical mechanics of a KP chain in an external force field   19
    2.2   The force-extension curve of the FJC   20
    2.3   The force-extension curve of the WLC   21
          2.3.1   Limiting form of the **U** matrix   21
          2.3.2   The analogy with the quantum rotator   23

2.4   The force-extension relationship of the discrete KP model   25
2.5   The DNA force-extension relationship   26
    2.5.1   Double-stranded DNA   26
    2.5.2   Single-stranded DNA   28

3.   DNA packaging and wrapping   29
    3.1   Packaging of genomic material in a DNA virus   29
        3.1.1   Shape and parameters   29
        3.1.2   Bending vs. hydration energies. Packaging force.
              Pressure on capsid walls   31
    3.2   Wrapping in nucleosomes   35
        3.2.1   Genomic packaging   35
        3.2.2   Bending energies at the nucleosome core   37
        3.2.3   Unwrapping the nucleosome   37

4.   Scattering from DNA in solution   39
    4.1   Elastic scattering from dilute solutions of macromolecules   39
    4.2   The structure factor of the Kratky–Porod chain   40
        4.2.1   An intermediate result   40
        4.2.2   Structure factor of the homogeneous KP chain . .   41
    4.3   Structure factor of the freely jointed chain (FJC)   41
    4.4   Structure factor of the WLC   43
        4.4.1   The approach to the continuum limit   43
        4.4.2   Numerical evaluation of the structure factor . . .   44
    4.5   Numerical evaluation of the end-to-end distance probabil-
        ity distribution function   45
    4.6   Structure factors of simple geometrical molecules   46
        4.6.1   Structure factor of a spherical molecule   47
        4.6.2   Structure factor of a cylindrical molecule   47
    4.7   Light scattering from long DNA molecules   49
    4.8   Small angle neutron scattering (SANS) from short DNA
        molecules   50

5.   Thermal unbinding of the double helix   53
    5.1   Introduction   53
        5.1.1   Discovery   53
        5.1.2   Basic Thermodynamics   55
    5.2   Base sequence and thermodynamic stability   57

		5.2.1	The nearest-neighbor model	57
		5.2.2	Enthalpies and entropies for neighboring base pairs	58
	5.3	Melting of longer DNA chains	61
		5.3.1	Internal vs. external melting	61
		5.3.2	Multistep melting	62
		5.3.3	Melting of long, synthetic, homogeneous duplexes	62
6.	**Mechanical unbinding of the double helix**			**65**
	6.1	Unzipping	65
		6.1.1	The experimental findings	65
		6.1.2	Mean-field theory	66
	6.2	Overstretching	68
		6.2.1	The overstretching transition	68
		6.2.2	Enthalpic corrections to the force-extension curve	69
		6.2.3	Force induced melting: the torsionally unconstrained case	70
		6.2.4	Torsionally constrained forced-induced melting . .	72
7.	**Helix-coil theory of DNA melting**			**77**
	7.1	Statistical models of helix-coil equilibria	77
		7.1.1	Synthetic polypeptides	77
		7.1.2	Helix growth vs. helix initiation	78
		7.1.3	The "all-or-nothing" (AON) model	79
		7.1.4	The zipper model	79
		7.1.5	The generalized zipper model	80
	7.2	Melting of infinitely long homogeneous DNA. The Poland–Scheraga model	82
		7.2.1	A useful shortcut	83
		7.2.2	Loop entropies	84
		7.2.3	The phase transition	84
		7.2.4	Bubbles and clusters	89
		7.2.5	A thermodynamic phase transition in a one-dimensional system?	90
8.	**Dynamical theory of DNA melting I. Fundamentals**			**93**
	8.1	Base pairs as dynamical entities	93
		8.1.1	Functional requirements	93
		8.1.2	Breathing of individual base pairs	93

8.1.3   Base pair lifetime   94
8.1.4   Low frequency vibrations   94
8.2   The homogeneous Peyrard–Bishop model   95
8.2.1   Definitions and Notation   95
8.2.2   Dynamics   96
8.2.3   Statistical mechanics   97
8.2.4   The phase transition   103
8.2.5   Bubbles   106
8.2.6   DNA unzipping in the PB model   108

9.   Dynamical theory of DNA melting II. Nonlinear stacking
interaction   109

9.1   The homogeneous Peyrard–Bishop–Dauxois (PBD) Hamiltonian   109
9.2   A local thermally activated barrier   111
9.3   Thermodynamic properties of the PBD Hamiltonian . . .   111
9.4   First-order transition: true or apparent?   114

10.   Dynamical Theory of DNA melting III: long, heterogeneous chains   117

10.1   Heterogeneity in the PBD model framework   117
10.1.1   Theory describes *internal* melting   117
10.1.2   The Hamiltonian and its model parameters   117
10.2   Statistical mechanics of the finite, heterogeneous chain . .   118
10.2.1   The partition function   118
10.2.2   The melting fraction   120
10.2.3   Computational issues   121
10.3   Computed melting profiles   122
10.3.1   The T-7 phage   122
10.3.2   The pBR322 plasmid   123
10.4   Predictive power of the PBD model   124
10.4.1   Dependence of the Morse depths on salt concentration   124
10.4.2   Parameter-free computation of melting profiles . .   125
10.4.3   Longer sequences   126
10.4.4   Unzipping   126

11.   Temperature dependent DNA flexibility   129

11.1 Introduction 129
11.2 Statistical mechanics of the heterogeneous KP chain . . . 130
    11.2.1 The partition function 130
    11.2.2 Correlations 130
    11.2.3 The second moment 131
11.3 Flexibility and melting 132
    11.3.1 Magnetic birefringence of DNA in solution 132
    11.3.2 Birefringence in the premelting and melting regime 133
11.4 Enhanced flexibility of a short sequence at elevated tem-
    peratures 136

12. Is DNA softer at the 100-nm scale?   139

12.1 Scattering from a heterogeneous Kratky–Porod chain . . . 139
    12.1.1 The probability distribution for the distance be-
        tween any two monomers 140
    12.1.2 The structure factor of the heterogeneous KP chain 141
12.2 Hinge-induced apparent softening of a 100-bp sequence . . 142
12.3 Permanent bends and effective softening of short sequences 144
12.4 Kinky DNA in solution 146
12.5 Concluding remarks 147

13. Thermodynamic stability of DNA hairpins   149

13.1 Self-complementary sequences 149
13.2 Biomolecular beacons 149
    13.2.1 Fluorescence resonance energy transfer (FRET)
        spectroscopy 150
    13.2.2 Fluorescence upon hybridization 150
13.3 Thermodynamic stability of hairpins 151
    13.3.1 Open-closed equilibria 151
    13.3.2 Hairpin statistical mechanics: stem vs loop 152
13.4 Excluded volume, SAWs and electrostatic repulsion in ss-
    DNA 154

Appendix A   Monte Carlo simulations of the Kratky–Porod chain   157

Appendix B   Landau's theorem on the absence of phase
transitions in one-dimensional systems   161

Appendix C   Dynamical theory of DNA melting: The soliton analogy 163

C.1   Soliton-like field configurations of the Peyrard–Bishop model 163
C.2   Thermal fluctuations and soliton stability 164

Appendix D   Numerical solution of the transfer integral equation   167

D.1   Gauss–Legendre quadratures 167
D.2   Application to the TI equation 168

Bibliography   169

Index   177

Chapter 1

Statistical mechanics of simple polymer chain models

1.1 Definitions and notation. Key quantities

Consider a chain of $N + 1$ monomers, consisting of N segments, each of length a. Any particular conformation of the chain can be described (cf. Fig.1.1) in terms of the direction vectors of all segments $\{\vec{t}_i, i = 1, \ldots, N\}$. The absolute positions of monomers in space $\{\vec{R}_i, i = 0, \ldots, N\}$ are given by

$$\vec{R}_i = a \sum_{j=1}^{i} \vec{t}_j, \quad i = 1, \ldots, N, \tag{1.1}$$

where $\vec{R}_0 = 0$ by convention.

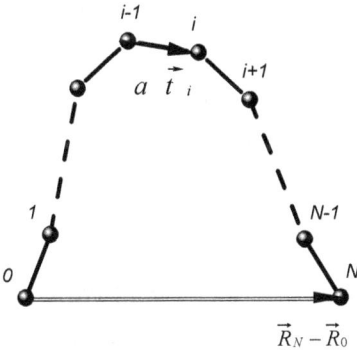

$$\vec{R}_N - \vec{R}_0$$

Fig. 1.1   Schematic representation of a random conformation of a polymer chain consisting of N segments of constant length a. The direction vector of the ith segment is $\vec{t}_i$, the position vector of the ith monomer is $\vec{R}_i$ and the end-to-end vector is $\vec{R}_N - \vec{R}_0$.

The reference length of the stretched chain, also known as its *contour length*, will be $L_0 = Na$. The actual end-to-end distance vector of the particular conformation will be equal to $\vec{R}_N$. Depending on the details of the conformation, the end-to-end distance $|\vec{R}_N|$ can take any value between 0 and L_0.

The effective size of a polymer in solution is determined as an average over all possible conformations, where each conformation contributes according to its probability of occurrence. In order to estimate this probability, it is necessary to have an idea of the energy connected with each particular conformation. Typically, a conformation may involve an energy cost because some monomers may be too close to each other. One then speaks of an *excluded volume effect*. Or, there may be a *rigidity* at the microscopic level which tends to enforce a more or less parallel orientation of neighboring segments.

1.1.1   *The freely jointed chain (FJC)*

In the absence of both of the above effects (excluded volume and/or local rigidity) randomness prevails. Does this mean that all conformations are equally probable? Not quite. The spatial structure of a polymer which is totally flexible at the local level — known as the *freely jointed chain (FJC)* — can be visualized as a *random walk* in three-dimensional space, where each step is of constant length and random direction. In other words, the end-to-end vector (i.e. the total vector displacement produced by the random walk) will be distributed according to a probability distribution

$$P(\vec{r}) = \frac{1}{(4\pi)^N} \int d\Omega_1 \cdots d\Omega_N \delta\left(\vec{r} - a\sum_{i=1}^{N} \vec{t_i}\right) \tag{1.2}$$

where the integrals are taken over possible orientations (polar and azimuthal) of each segment vector.

I will evaluate the integral (1.2) later.

1.1.2   *Local stiffness. Elastic energy. The wormlike chain*

Let me now consider the case where, in the absence of any excluded volume effects, the polymer is exhibits a finite local rigidity. This favors a parallel orientation of neighboring segments. Conversely, the total cost of nonparallel orientation of all neighboring segments in the chain is a measure of the total elastic energy. One of the simplest approaches to describe and parametrize the energy associated with polymer flexibility is in terms of the Kratky–Porod model [Kratky and Porod (1949)]:

$$H_0 = -\frac{\kappa}{a} \sum_{j=1}^{N-1} (\vec{t}_j \cdot \vec{t}_{j+1} - 1) \tag{1.3}$$

$$\rightarrow \frac{\kappa}{2} \int_0^{L_0} ds \left| \frac{\partial \vec{t}}{\partial s} \right|^2, \tag{1.4}$$

where κ is a measure of the chain's bending stiffness and the second line represents the continuum limit (1.4) as $N \to \infty, a \to 0, Na = L_0$. Note that in the latter case the local orientation of the chain is now described by the continuum unit vector $\vec{t}(s)$ which is locally tangent to the chain, and the local energy density of any particular conformation is determined by the local curvature $\left| \frac{\partial \vec{t}}{\partial s} \right|$ and the stiffness. The continuum version of the Kratky–Porod model is known as the *wormlike chain* (WLC) model.

In the following, I will derive some key properties of the discrete Kratky–Porod model [1] and show how these reduce in the continuum limiting case of the WLC.

1.2 Statistical mechanics of the Kratky–Porod (KP) chain

1.2.1 *The partition function*

At any given temperature T, the canonical partition function of the chain (1.3) with free ends

$$Z_N \equiv \int d\Omega_1 \cdots d\Omega_N \, e^{-\beta H_0} = \int d\Omega_1 \cdots d\Omega_N \prod_{j=1}^{N-1} e^{b(\vec{t}_j \cdot \vec{t}_{j+1} - 1)}, \tag{1.5}$$

where $b = \beta \kappa / a$, $\beta = 1/(k_B T)$ and k_B is the Boltzmann constant, can be evaluated by successive integrations over all orientation angles $d\Omega_j \equiv d\phi_j d\theta_j \sin \theta_j$ of the direction vectors $\vec{t}_j$. This can be done because (1.3) is isotropic and any unit vector $\vec{t}$ has the property

$$\int d\Omega e^{b\vec{t} \cdot \vec{t}'} \equiv \int_0^{2\pi} d\phi \int_0^\pi d\theta \sin \theta e^{b \cos \theta} = 4\pi i_0(b) \quad \forall \vec{t}', \tag{1.6}$$

where $i_0(b) = \sinh b / b$ is the modified spherical Bessel function of zeroth order. Repeated use of (1.6) results in the partition function

$$Z_N = 4\pi \left[4\pi e^{-b} i_0(b) \right]^{N-1}, \tag{1.7}$$

where the extra 4π factor comes from the last integration.

---

[1]Note that the discrete Kratky–Porod model is mathematically equivalent to the classical limit of the Heisenberg ferromagnetic chain.

1.2.1.1 *The limit of the freely jointed chain*

In the limiting case $b \to 0$ (1.7) reduces to the partition function of the FJC,

$$Z_N = (4\pi)^N, \tag{1.8}$$

which is simply a measure of the available phase space.

1.2.2 *Correlations. Persistence length*

Correlations of the orientation of the rth and sth segments are defined as

$$<\vec{t}_r \cdot \vec{t}_s> = \frac{1}{Z_N} \int d\Omega_1 \cdots d\Omega_N \, \vec{t}_r \cdot \vec{t}_s \prod_{j=1}^{N-1} e^{b(\vec{t}_j \cdot \vec{t}_{j+1} - 1)}, \tag{1.9}$$

and can be calculated [Fisher (1964)] from the general property of unit vectors

$$\int d\Omega e^{b\vec{t} \cdot \vec{t}'} \vec{t} = 4\pi i_1(b)\vec{t}', \tag{1.10}$$

where $i_1(b) = (b \cosh b - \sinh b)/b^2$ is the modified spherical Bessel function of first order.

Repeated application of (1.10) in (1.9) results in

$$<\vec{t}_r \cdot \vec{t}_s> = \left(\frac{i_1(b)}{i_0(b)}\right)^{|r-s|} \equiv e^{-|r-s|a/\lambda}, \tag{1.11}$$

where I identify the *persistence length* λ as the scale over which orientational correlations decay. Inverting (1.11) results in

$$\frac{a}{\lambda} = -\ln[i_1(b)/i_0(b)] = -\ln\left(\coth b - \frac{1}{b}\right), \tag{1.12}$$

or,

$$\lambda = \frac{\kappa}{k_B T} \tag{1.13}$$

in the continuum (WLC) limit.

1.2.2.1 *The limit of the freely jointed chain*

In the limit $b \to 0$ (1.12) gives $\lambda = 0$. This expresses the complete lack of correlations

$$<\vec{t}_r \cdot \vec{t}_s> = <\vec{t}_r> \cdot <\vec{t}_s> = 0 \quad \forall r \neq s. \tag{1.14}$$

1.2.3 Moments

It is possible to calculate directly a few low-order moments of the end-to-end distance. For example

$$< R_N^2 > \equiv \left\langle \left(\sum_{j=1}^{N} \vec{t}_j \right)^2 \right\rangle a^2 = \sum_{j=1}^{N} \sum_{i=1}^{N} < \vec{t}_j \cdot \vec{t}_i > a^2 \qquad (1.15)$$

can be computed using (1.11). The result is

$$< R_N^2 > = \frac{1}{1 - \mu} \left[(1 + \mu)N - 2\mu \frac{1 - \mu^N}{1 - \mu} \right] a^2, \qquad (1.16)$$

where

$$\mu = e^{-a/\lambda} = \coth b - 1/b. \qquad (1.17)$$

Interesting limiting cases are:

- the freely jointed chain,

$$< R_N^2 > = Na^2$$

- the continuum limit (WLC). In this case $\mu = e^{-a/\lambda}$ approaches unity and care must be taken in obtaining the limits. The point to recognize is that in this limit $\mu^N \to e^{-L_0/\lambda}$ and $1 - \mu \to a/\lambda$. The result in the continuum limit is therefore

$$< R_L^2 > = 2\lambda L_0 - 2\lambda^2 (1 - e^{-L_0/\lambda}). \qquad (1.18)$$

1.2.4 Radius of gyration

A better measure of the physical size of a polymer than the average end-to-end distance is the radius of gyration R_g defined as

$$R_g^2 = \frac{1}{N + 1} \sum_{i=0}^{N} \left\langle (\vec{R}_i - \vec{R}_{CM})^2 \right\rangle \qquad (1.19)$$

where

$$\vec{R}_{CM} = \frac{1}{N + 1} \sum_{i=0}^{N} \vec{R}_i \qquad (1.20)$$

is the position vector of the chain's center of mass. It is straightforward to show, using (1.20) and (1.19) that

$$R_g^2 = \frac{1}{2(N + 1)^2} \sum_{m=0}^{N} \sum_{n=0}^{N} \left\langle (\vec{R}_m - \vec{R}_n)^2 \right\rangle. \qquad (1.21)$$

The radius of gyration is a quantity directly accessible in a variety of scattering experiments, e.g. using visible light, x-rays or neutrons, depending on the size of the macromolecule to be studied. For the particular case of the Kratky–Porod model, it can be calculated exactly as follows. Noting the property

$$\vec{R}_m - \vec{R}_n = a \sum_{j=m}^{n-1} \vec{t}_j, \quad m < n \tag{1.22}$$

and making use of (1.11) we obtain

$$< (\vec{R}_m - \vec{R}_n)^2 > = \frac{1+\mu}{1-\mu}(n-m)a^2 - \frac{2\mu}{(1-\mu)^2}(1-\mu^{n-m})a^2, \quad m < n \tag{1.23}$$

where $\mu = \exp(-a/\lambda)$. Using (1.23) in (1.21) leads to

$$\left(\frac{R_g}{L_0}\right)^2 = \frac{1}{6} \cdot \frac{1+\mu}{1-\mu} \cdot \frac{(N+2)}{N(N+1)} - \frac{\mu}{(1-\mu)^2} \cdot \frac{1}{N(N+1)}$$
$$+ 2\frac{\mu^2}{(1-\mu)^3} \cdot \frac{1}{N(N+1)^2}$$
$$- 2\frac{\mu^3}{(1-\mu)^4} \cdot \frac{1}{[N(N+1)]^2} \cdot (1-\mu^N) \tag{1.24}$$

or, taking again the continuum limit,

$$\left(\frac{R_g}{L_0}\right)^2 = \frac{1}{3}\frac{\lambda}{L_0} - \left(\frac{\lambda}{L_0}\right)^2 + 2\left(\frac{\lambda}{L_0}\right)^3 - 2\left(\frac{\lambda}{L_0}\right)^4 \cdot (1 - e^{-L_0/\lambda}) \tag{1.25}$$

for the WLC [Benoit and Doty (1953)].

1.2.4.1   *The limit of the freely jointed chain*

In the limit $b \to 0$, $r \to 0$ and (1.24) reduces to

$$R_g^2 = \frac{1}{6} \cdot \frac{N+2}{N+1} \cdot Na^2, \tag{1.26}$$

and if, in addition, $N \gg 1$,

$$R_g^2 = \frac{1}{6} \cdot Na^2. \tag{1.27}$$

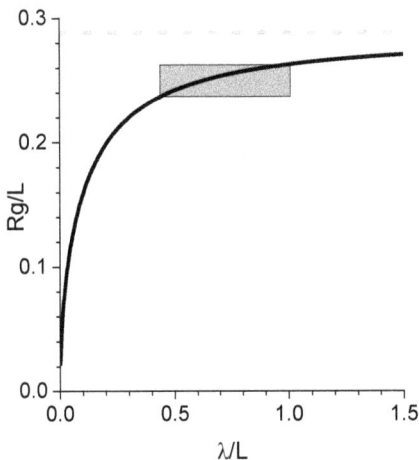

Fig. 1.2   The Benoit–Doty relationship (1.25) beween R_g, L_0 and λ for the WLC. Also shown (dotted line) is the "hard rod" limit, $\lambda/L_0 \to \infty$, $R_g/L_0 \to \sqrt{1/12}$. Note that as the chain gets stiffer, it becomes increasingly difficult to use measured values of R_g in order to estimate λ. The shaded box demonstrates that, at $R_g/L_0 \sim .25$, a $\pm 5\%$ error in R_g translates to more than a factor of 2 uncertainty in λ.

1.3   The end-to-end distance probability distribution

The end-to-end distance probability distribution is defined as a canonical ensemble average over all possible conformations of the polymer chain

$$P_N(\vec{r}) = \left\langle \delta \left[\vec{r} - (\vec{R}_N - \vec{R}_0) \right] \right\rangle$$

$$= \frac{1}{Z_N} \int d\Omega_1 \cdots d\Omega_N \prod_{j=1}^{N-1} e^{b(\vec{t}_j \cdot \vec{t}_{j+1} - 1)} \delta \left(\vec{r} - a \sum_{i=1}^{N} \vec{t}_i \right) \quad (1.28)$$

Eq. (1.28) generalizes (1.2) for systems with nonzero stiffness.

In order to make progress with the computation of the end-to-end distance distribution it will be necessary to use the Fourier representation of the Dirac delta function

$$\delta \left(\vec{r} - a \sum_{i=1}^{N} \vec{t}_i \right) = \frac{1}{(2\pi)^3} \int d\vec{q} \, e^{i\vec{q} \cdot (\vec{r} - a \sum_{j=1}^{N} \vec{t}_j)}. \quad (1.29)$$

We then recognize that the Fourier transform of $P_N(\vec{r})$ can be written in the form

$$\hat{P}_N(\vec{q}) = \frac{1}{Z_N} \int d\Omega_1 \cdots d\Omega_N \prod_{j=1}^{N-1} e^{b(\vec{t}_j \cdot \vec{t}_{j+1} - 1)} \prod_{j=1}^{N} e^{-ia\vec{q} \cdot \vec{t}_j}, \tag{1.30}$$

which allows us to exploit the rotational invariance of the interactions. I will use a fixed system of spherical coordinates, taking the z-axis parallel to the $\vec{q}$ vector; as a consequence, the $\vec{q}$-dependent term does not enter the integration over azimuthal angles. I introduce the standard expansion of the isotropic interaction in terms of spherical harmonics

$$e^{b\vec{t}_j \cdot \vec{t}_{j+1}} = \sum_{l=0}^{\infty} \sum_{m=-l}^{l} i_l(b) \, Y_{lm}(\Omega_j) Y_{lm}^*(\Omega_{j+1}) \tag{1.31}$$

where i_l is the modified spherical Bessel function of lth order,

$$Y_{lm}(\Omega) = \left[\frac{2l+1}{4\pi} \frac{(l+m)!}{(l-m)!} \right]^{1/2} P_l^m(\cos\theta) e^{im\phi} \tag{1.32}$$

and P_l^m is the associated Legendre polynomial. I now note that the integrations over azimuthal angles generate the contributions

$$\int_0^{2\pi} d\phi_j \, e^{i\phi_j(m_j - m_{j-1})} = 2\pi \delta_{m_j, m_{j-1}}, \quad j = 2, \ldots, N-1,$$

$$\int_0^{2\pi} d\phi_1 \, e^{i\phi_1 m_1} = 2\pi \delta_{m_1, 0},$$

$$\int_0^{2\pi} d\phi_N \, e^{i\phi_1 m_N} = 2\pi \delta_{m_N, 0},$$

i.e. only the $m=0$ terms of the expansion (1.31) contribute, so that

$$\hat{P}_N(\vec{q}) = \frac{1}{2^N} \sum_{l_1, l_2, \cdots, l_{N-1}} \prod_{j=1}^{N-1} [(2l_j + 1)\hat{i}_{l_j}(b)] \int_{-1}^{1} d\mu_1 d\mu_2 \cdots d\mu_N P_{l_1}(\mu_1)$$

$$\cdot P_{l_1}(\mu_2) P_{l_2}(\mu_2) P_{l_2}(\mu_3) \cdots P_{l_{N-1}}(\mu_{N-1}) P_{l_{N-1}}(\mu_N)$$

$$\cdot e^{-iqa\mu_1} \cdots e^{-iqa\mu_N}, \tag{1.33}$$

where $P_l{}'$ is the Legendre polynomial of order l and I have absorbed the $i_0(b)$ factors of the partition function by defining $\hat{i}_l(b) \equiv i_l(b)/i_0(b)$. Note further that there is no dependence on the orientation of the vector $\vec{q}$. Owing to the spatial isotropy of the interactions, the end-to-end distribution function depends only on $r = |\vec{r}|$ and its Fourier transform on $q = |\vec{q}|$.

Defining the symmetric matrix $\mathbf{F}$ with elements

$$F_{ll'} \equiv \frac{1}{2} [\hat{i}_l(b)] \hat{i}_{l'}(b) (2l+1)(2l'+1)]^{1/2} f_{ll'}(q) \tag{1.34}$$

where

$$f_{ll'}(qa) \equiv \int_{-1}^{1} d\mu P_l(\mu) P_{l'}(\mu) e^{-iqa\mu} \tag{1.35}$$

and, noting that $\hat{i}_0(b) = 1 \; \forall b$, I can cast (1.33) in the more compact form

$$\hat{P}_N(\vec{q}) = (\mathbf{F}^N)_{00}. \tag{1.36}$$

1.3.0.1 *Closed form expressions for the $f_{ll'}$*

It is possible to express the integral (1.35) in closed form by making use of (i) the expansion

$$e^{-ix\mu} = \sum_{k=0}^{\infty} (2k+1)(-i)^k j_k(x) P_k(\mu) \tag{1.37}$$

where the j's are Bessel functions, e.g. $j_0(x) = \sin x/x$, and (ii) the integral formula for the product of three Legendre polynomials [Gradshteyn and Ryzhik (2007)]

$$\int_{-1}^{1} d\mu P_k(\mu) P_l(\mu) P_{l'}(\mu) = \begin{cases} \frac{1}{r+\frac{1}{2}} \frac{\Psi(r-k)\Psi(r-l)\Psi(r-l')}{\Psi(r)} & \text{if } k+l+l' = 2r \\ & \text{and } |l'-l| \le k, \\ 0 & \text{otherwise,} \end{cases}$$

where

$$\Psi(n) = \frac{\Gamma(n+\frac{1}{2})}{\Gamma(n+1)\Gamma(\frac{1}{2})} = \prod_{j=1}^{n} \left(1 - \frac{1}{2j}\right). \tag{1.38}$$

The result is

$$f_{ll'}(qa) = \sum_{k=|l-l'|, k+l+l'=2r}^{l+l'} (2k+1)(-i)^k \frac{1}{r+1/2}$$
$$\cdot \frac{\Psi(r-k)\Psi(r-l)\Psi(r-l')}{\Psi(r)} j_k(qa). \tag{1.39}$$

Note that $f_{ll'}$ is either real (if l and l' are both even or odd) or pure imaginary (if one of l and l' is even and the other odd). In the first case the sum (1.39) contains only even, in the second case only odd terms.

1.3.0.2 *The limit of the freely jointed chain*

In the limit of the freely jointed chain, $\kappa = 0, b = 0$ all the $\hat{i}_l$s vanish, except for $\hat{i}_0 = 1$. The matrix F collapses to a scalar, equal to $f_{00} = j_0(qa)$, i.e.

$$\hat{P}_N(\vec{q}) = \left(\frac{\sin qa}{qa} \right)^N . \tag{1.40}$$

It is possible to invert the Fourier transform

$$P_N(\vec{r}) = \int \frac{d\vec{q}}{(2\pi)^3} e^{i\vec{q}\cdot\vec{r}} \left(\frac{\sin qa}{qa} \right)^N , \tag{1.41}$$

and obtain [Gradshteyn and Ryzhik (2007)]

$$P_N(r) = \frac{1}{2\pi^2 r a^2} \int_0^\infty dx\, x \sin(\frac{r}{a}x) \left(\frac{\sin x}{x} \right)^N$$
$$= \begin{cases} \frac{1}{2^{N+1}(N-2)!\pi a^2 r} \sum_{k=0}^{[s]} (-1)^k \binom{N}{k} (N - \frac{r}{a} - 2k)^{N-2} & \text{if } 0 < r < Na \\ 0 & \text{if } r \geq Na \end{cases}$$
$$\tag{1.42}$$

where $r = |\vec{r}|$, $s = (N - r/a)/2$ and $[s]$ is the integer part of s.

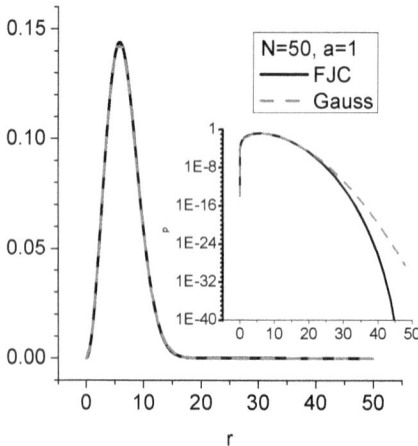

Fig. 1.3 Comparison of the exact FJC end-to-end distance distribution function $p_N(r) = 4\pi r^2 P_N(r)$ with its Gaussian approximation (1.44). Both distributions are also plotted in the inset in semilog scale, in order to emphasize behavior at the short- and long-distance tails.

(1.42), although exact, is not particularly informative. However, its limiting form as N becomes large is particularly simple. Going back to (1.41), and recognizing that contributions to the integral can only come from very small values of qa, we use the expansion [Abramowitz and Stegun (1964)]

$$\ln\left(\frac{\sin x}{x}\right) = \sum_{l=1}^{\infty}(-1)^l \frac{B_{2l}2^{2l}}{(2l)!2l}x^{2l}, \quad |x| < \pi, \tag{1.43}$$

where $B_2 = 1/6$, $B_4 = -1/30$ are the Bernoulli numbers. To leading order,

$$\left(\frac{\sin qa}{qa}\right)^N \approx e^{-N(qa)^2/6}$$

which permits the factorization of the integral (1.41) in Cartesian coordinates and results in

$$P_N(\vec{r}) \approx \left(\frac{3}{2\pi Na^2}\right)^{3/2} e^{-\frac{3}{2}\cdot\frac{r^2}{Na^2}}, \tag{1.44}$$

which coincides with the end-to-end distance distribution function of the *Gaussian chain* (cf. next section). Fig. 1.3 illustrates the quality of the approximation for $N = 50$. The function plotted is, in both cases, the end-to-end distance distribution after integrating over the angular variables, i.e. $p_N(r) = 4\pi r^2 P_N(r)$.

1.4 The Gaussian chain

Suppose we relax the constraint imposed in 1.1 that all segments should be of identical length. Instead, let the length of each segment vector $\vec{\xi}_n = \vec{R}_n - \vec{R}_{n-1}$ be randomly distributed, according to a Gaussian distribution with a width a. Such a distribution will arise at a given temperature T if the total energy is

$$H_G = \frac{3}{2}\frac{k_B T}{a^2}\xi_n^2. \tag{1.45}$$

The canonical average of any function of the monomer distances $A(\{\vec{R}_n\})$ can be calculated in terms of the unconstrained integral

$$< A(\{\vec{R}_n\}) > = \frac{1}{Z}\int d\vec{R}_1 \cdots d\vec{R}_N \, A(\{\vec{R}_n\}) \, e^{-H_G/k_B T}, \tag{1.46}$$

where

$$Z = \int d\vec{R}_1 \cdots d\vec{R}_N \, e^{-H_G/k_B T}$$

$$= \int d\vec{\xi}_1 \cdots d\vec{\xi}_N \prod_{i=1}^{N} e^{-\frac{3}{2a^2}\vec{\xi}_i^2}$$

$$= \left(\frac{2\pi a^2}{3}\right)^{3N/2} \tag{1.47}$$

is the partition function. Functions of interest include

- the probability distribution of the end-to-end distance vector

$$P_N^{(G)}(r) = \left(\frac{3}{2\pi N a^2}\right)^{3/2} e^{-\frac{3}{2}\frac{r^2}{N a^2}}, \tag{1.48}$$

- the radius of gyration (defined in 1.2.4)

$$R_g^2 = \frac{1}{6}N a^2, \tag{1.49}$$

- moments of the end-to-end distance distribution

$$< r^{2n} > = \int_{-\infty}^{\infty} dr \, r^{2n} \, P_N^{(G)}(r) \tag{1.50}$$

e.g. $< r^2 > = N a^2$. Note that, owing to the Gaussian integrations involved, all moments can be calculated straightforwardly.

A comparison of the end-to-end distance distribution functions between the purely Gaussian polymer chain and its FJC counterpart can be visualized in Fig. 1.3.

1.4.0.1   *Numerical results for $P_N(r)$*

Given the expression (1.36) for the Fourier transform of the end-to-end distribution function $P_N(r)$ and the explicit form of the matrix elements (1.34) and (1.39) we can in principle compute the Fourier transform and invert it to obtain the distribution in real space. An efficient numerical method for doing this in the WLC limit is described in 4.5. Results for chains of varying flexibility are shown in Fig. 1.4.

An alternative, available even in cases where the model does not allow analytical or numerical computations is to perform a Monte–Carlo (MC) simulation of the chain. A brief description of the method is provided in Appendix A. Fig. 1.4 includes results of such a calculation for a stiff chain $(\lambda/L_0 = 1.0)$.

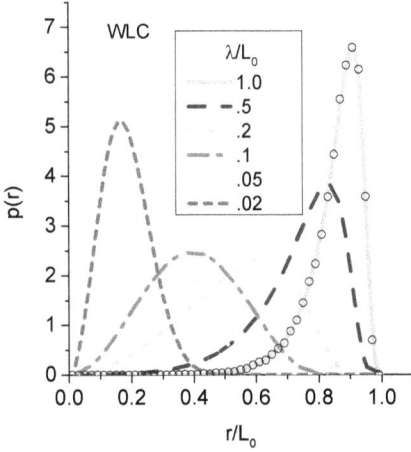

Fig. 1.4 The end-to-end distance probability distribution of the wormlike chain for varying values of the dimensionless flexibility parameter, ranging from very flexible $\lambda/L_0 = 0.2$, to stiff $\lambda/L_0 = 1.0$. The points refer to the latter, stiff case and have been obtained by a Monte Carlo (MC) calculation.

1.4.0.2   *The probability of loop formation*

In the limit $r \to 0$, the function $P_N(\vec{r})$ expresses the probability that the ends of an N-segment polymer will meet. Using the leading correction of (1.43) one obtains

$$P_N(0) = \left(\frac{3}{2\pi N a^2} \right)^{3/2} \left[1 - \frac{3}{4} \cdot \frac{1}{N} + \mathcal{O}(\frac{1}{N^2}) \right] \qquad (1.51)$$

for the probability of ring closure (spontaneous loop formation) in the FJC model.

In the case of the WLC chain, the probability of ring closure can be computed numerically. The result depends on the chain length via the dimensionless parameter L_0/λ and is shown graphically in Fig. 1.5. Ring closure is most probable around $L_0/\lambda = 3.38$. As the length increases, the result scales as $L_0^{-3/2}$ and becomes indistinguishable from the FJC limit (1.51), with $N = L_0/(2\lambda)$.

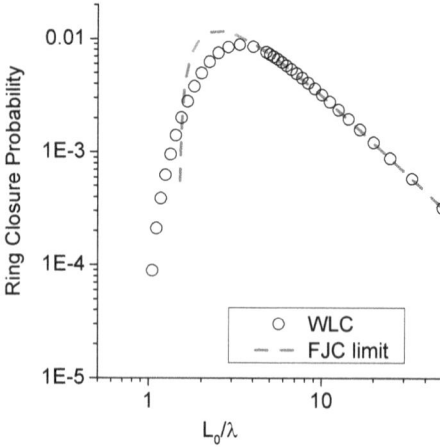

Fig. 1.5 The probability of ring closure as a function of the relative flexibility. The points denote numerical computations for the WLC model. For comparison, I also plot the function (1.51) with $N = L_0/(2\lambda)$ (FJC limit).

1.5 Excluded volume effects

1.5.1 *Characterization of the end-to-end distance distribution function*

A linear polymer chain constructed as a succession of randomly oriented segments (FJC or Gaussian chain) is an idealization which ignores fundamental spatial restrictions arising from the fact that two different monomers cannot simultaneously occupy the same position in space. This "excluded volume effect" can be mathematically expressed in terms of the end-to-end distance probability distribution function [Fisher (1966b)]

$$P_N(r) = \frac{C}{N^{d\nu}} f\left(\frac{r}{N^\nu}\right) \qquad (1.52)$$

where d is the dimensionality of space, ν is the correlation length exponent, C some numerical normalization constant and, for the sake of simplicity, lengths are measured in units of the monomer distance. The function f is characterized by its behavior at short

$$f(x) \propto x^\theta, \quad x \ll 1,$$

and at long

$$f(x) \propto e^{-x^\delta}, \quad x \gg 1,$$

distances through the critical exponents θ and δ. The three exponents are not all independent of each other. δ and ν satisfy the scaling relationship [Fisher (1966b)]

$$\delta = \frac{1}{1 - \nu}.$$

1.5.2   *Critical exponents and observable quantities*

Before discussing the numerical values of these exponents, let me relate them to observable quantities. The second moment of the distribution, closely related to the square of the radius of gyration, satisfies

$$< r^2 > \propto R_g^2 \propto N^{2\nu}.$$

This implies a characteristic length scale $R_g \propto N^\nu$.

The probability of ring closure (a.k.a. loop formation, cyclization) can be defined as

$$p_c(N) \propto \int_0^b dr \, r^{d-1} P_N(r) \propto \frac{1}{N^c},$$

where

$$c = (d + \theta)\nu, \tag{1.53}$$

i.e. as the probability that the end segment of the chain will be within a sphere of radius b, typically equal or smaller than the monomer distance a, and is controlled by both θ and ν.

In the Gaussian polymer chain, which can be thought of as random walk in space, $\nu = 1/2$, $\theta = 0$. $\delta = 2$. The distribution function (1.44) approaches a constant in the limit $r \to 0$. The quantity p_c decays as $N^{-3/2}$ in three dimensions. The same is true for the FJC chain.

1.5.3   *Values of the critical exponents. Relationship to the self-avoiding walk problem*

A number of problems in polymer physics can be successfully modeled by statistical walks on a lattice. A prime example is the *self-avoiding walk* (SAW) in which a walker may successively occupy any neighboring lattice sites that have not been previously visited. The spatial constraints imposed by this rule turn out to be equivalent to those of the excluded volume problem in polymer statistics. They lead to vanishing probability distribution at short distances, i.e. $\theta > 0$, and to systematically larger values of the mean square displacement, as compared to the purely random walk, i.e. to values $\nu > 1/2$. Historically, enumerations of polygons on a lattice provided a computationally efficient way to obtain information not be available via a realistic, off-lattice Monte–Carlo simulation of a polymer system. Polygon enumeration analysis of a SAW on a 3-dimensional face-centered lattice, [Martin *et al.* (1967)] led to the following results:

- The total number of walks with N steps starting at the origin, u_N, varies as

$$u_N = A\bar{z}^N\, N^{\gamma-1} \qquad (1.54)$$

  where A is a numerical constant of order unity, $\bar{z} = 10.035$, and $\gamma = 1.166$. The exponent α is "universal", in the sense that is the same for all 3-dimensional lattices studied — whereas the constants $\bar{z}$ and A depend on the type of lattice — and was found to be related to the exponent θ via the scaling relationship[des Cloizeaux (1974)]

$$\gamma - 1 = \theta\nu. \qquad (1.55)$$

- The number of walks starting at the origin and returning there after N steps, c_N, varies, in the limit of large N, as

$$c_N = A'\bar{z}^N\, N^{\beta} \qquad (1.56)$$

  where A' is another numerical constant of order unity, and $\beta = -1.75$. Again, the exponent β is related to ν a scaling relationship,

$$\beta = -d\nu. \qquad (1.57)$$

Consequently,

- the probability of ring closure in a 3-d lattice SAW will depend on the number of steps as

$$p_c(N) \equiv \frac{c_N}{u_N} \propto N^{-c} \qquad (1.58)$$

with $c = -\beta + \gamma - 1 = (d + \theta)\nu = 1.92$.

Direct MC simulations of the SAW on a cubic lattice [Dayantis and Palierne (1991)] determined a value $\nu = 0.5919$ and $\theta = 0.27$, leading to $c = 1.935$. Finally, renormalization group calculations [des Cloizeaux, J. (1980)] provide consensus values $\nu = 0.588$ and $\theta = 0.273$, implying

$$p_c(N) \propto N^{-1.925}.$$

It is also possible — although computationally prohibitive for very large systems — to perform a direct MC calculation for an FJC chain with the added constraint of excluded volume. In such a calculation, each monomer claims for itself a sphere of diameter D less or equal to the segment length a. This is equivalent to a pair interaction with a hard-core potential energy acting between all segment pairs in the chain and leads to a high number of rejected conformations.

Fig. 1.6 summarizes the results of such a direct MC calculation on a chain with $N = 30$ segments, $D/a = 1$ and 2×10^6 accepted conformations. The left panel shows the end-to-end distance probability distribution, from which a value of $\theta = 0.29$ can be extracted. The right panel shows the results of the same type of MC calculation for the radius of gyration, with an estimated value $\nu = 0.6228$. The values of these exponents are close (although not identical) to the consensus asymptotic values.

Fig 13.3 (left panel) displays the results of a similar MC calculation of the cyclization rate as a function of the number of segments $6 < N < 40$, in the case an FJC chain with the same excluded volume parameter. The power law exponent extracted from the data, 1.926 ± 0.10 corresponds almost exactly with the consensus asymptotic values.

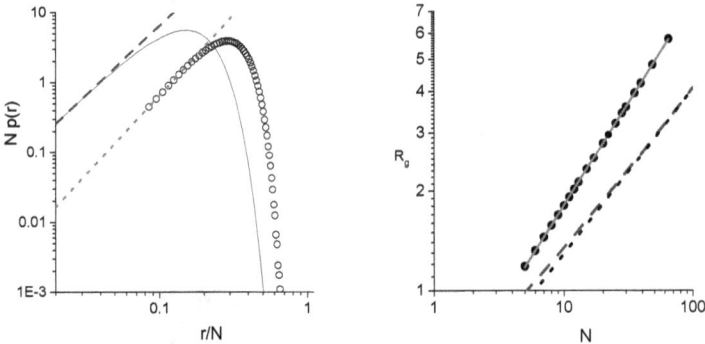

Fig. 1.6 *Left panel:* The end-to-end distance probability distribution $p_N(r)$ for an FJC chain with $N = 30$ and an excluded volume parameter $D/a = 1$. The circles denote the results of an MC calculation with 2×10^6 accepted conformations. The dotted line, obtained by a least squares fit of the lowest 6 points has a slope $2 + \theta = 2.29$. Also shown, for comparison, is the pure FJC result (1.42) with the asymptotic slope (dashed line) equal to 2. *Right panel:* The radius of gyration, in units of the segment length, as obtained from an MC calculation (full circles) with the same parameters. The straight line is a least squares fit with a slope $\nu = 0.6228$. Also shown (dashed line) is the FJC result (1.26) and the Gaussian limit (1.49) (dotted line).

Chapter 2

Entropic elasticity: the DNA force-extension relationship

The elastic response of common polymeric materials, e.g. rubber, can be traced to *entropic* rather than *enthalpic* changes of their state[Flory (1953)]. A long, flexible polymer coiling randomly in solution will have a typical end-to-end distance far smaller than its contour length. Consequently, the work done by a force acting to stretch it will be mostly spent in changing the spatial conformation of the molecule; chemical bonding, and thus internal enthalpies, play a minor role.

2.1 Statistical mechanics of a KP chain in an external force field

Consider, as is the case in various single-molecule experiments, an external force $\vec{f}$ acting on the end monomer of a Kratky–Porod polymeric chain, while the first is held fixed. This adds an extra term

$$-\vec{f} \cdot (\vec{R}_N - \vec{R}_0)$$

to the total energy (1.3), i.e. now

$$H = -\frac{\kappa}{a} \sum_{j=1}^{N-1} (\vec{t}_j \cdot \vec{t}_{j+1} - 1) - a\vec{f} \cdot \sum_{j=1}^{N} \vec{t}_j \qquad (2.1)$$

$$\rightarrow \frac{\kappa}{2} \int_0^{L_0} ds \left| \frac{\partial \vec{t}}{\partial s} \right|^2 - \vec{f} \cdot \int_0^L ds\, \vec{t}(s), \qquad (2.2)$$

where the second line refers to the continuum limit. The canonical partition function in the presence of the external force is

$$Z_N(f) = \int d\Omega_1 \cdots d\Omega_N \prod_{j=1}^{N-1} e^{b(\vec{t}_j \cdot \vec{t}_{j+1} - 1)} \prod_{j=1}^{N} e^{\beta a \vec{f} \cdot \vec{t}_j}, \qquad (2.3)$$

and the quantity of interest is the average extension

$$< \vec{R}_N - \vec{R}_0 > = a \sum_{j=1}^{N} < \vec{t}_j > = \frac{1}{\beta} \frac{\partial}{\partial f} \ln Z_N(f). \tag{2.4}$$

Before proceeding with the calculation of the average extension as a function of the applied force, let me make two comments.

First, as a consequence of the isotropy of the energy H_0, the average extension will always be in the direction of the external force. In practice therefore, I will be calculating the quantity

$$L = < (\vec{R}_N - \vec{R}_0) \cdot \hat{f} >$$

i.e. the component of the extension along the direction of the force.

Second, since I am computing the derivative of $\ln Z_N(f)$ with respect to the force, it is always possible to divide $Z_N(f)$ by a normalizing force-independent factor without changing the extension. I will use Z_N, the force-free partition function 1.7 as such a factor. This means that the key quantity to be computed is

$$\hat{Z}_N(f) = \frac{Z_N(f)}{Z_N}. \tag{2.5}$$

Note now that the expressions (1.30) and (2.5) are essentially identical, except for the substitution $i\vec{q} \to \beta\vec{f}$. Therefore, by direct analogy with (1.36), I can recast $\hat{Z}_N(f)$ in matrix form,

$$\hat{Z}_N(f) = (\mathbf{U}^N)_{00}, \tag{2.6}$$

where the elements of the matrix are given by

$$U_{ll'} \equiv \frac{1}{2} [\hat{i}_l(b)\hat{i}_{l'}(b)(2l+1)(2l'+1)]^{1/2} \int_{-1}^{1} dx P_l(x) P_{l'}(x) e^{\beta a f x}. \tag{2.7}$$

The above equations provide a basis for a full numerical calculation of the force-extension curve in the case of discrete KP chains. However, if taking the continuum limit can be justified on specific physical grounds (e.g. long DNA chains with tens of thousands of base pairs), one can do significantly better. Before I sketch this approach, let me look at a mathematically simpler case.

2.2 The force-extension curve of the FJC

In the limiting case of the freely jointed chain the partition function (2.3) reduces to a product of N independent integrals, each equal to (cf. (1.6)) $4\pi i_0(\beta a f)$. This leads to

$$\hat{Z}_N(f) = [i_0(\beta a f)]^N$$

and, using (2.4)

$$\frac{L}{L_0} = \frac{i_0'(\beta a f)}{i_0(\beta a f)} = \coth(\beta a f) - \frac{1}{\beta a f}. \tag{2.8}$$

At low external force fields, this reduces to

$$L/L_0 \sim (a/3k_B T)f \quad f \ll k_B T/a.$$

The chain responds linearly to the external force with an effective stiffness constant $a/3k_B T$.

At high forces

$$L/L_0 \sim 1 - k_B T/(af) \quad f \gg k_B T/a,$$

the extension approaches the fully stretched length. The FJC model is sometimes useful in "mimicking" the behavior of a stiffer WLC chain. Considering an FJC chain with an *effective* monomer distance equal to twice the persistence length (known as the *Kuhn length* of the polymer); the *effective* number of monomers is taken to be such that the contour length is equal to that of the WLC chain. The force extension curve at low forces will then be identical to that of the underlying WLC (cf. Fig. 2.1).

2.3 The force-extension curve of the WLC

2.3.1 *Limiting form of the* U *matrix*

In order to perform the continuum limit of the exponentiated form (2.6) I keep terms of order $a = L_0/N$ in (2.7). These come from two sources:

- expanding the exponential $e^{\beta a f x} \approx 1 + a\beta f x$ generates - besides the diagonal term of zeroth order in a coming from the unity - a contribution proportional to

$$\int_{-1}^{1} dx\, P_l(x) P_{l'}(x) x = \frac{2}{(2l+1)(2l'+1)} \{l'\delta_{l',l+1} + l\delta_{l',l-1}\}$$

- using the asymptotic forms of the modified spherical Bessel functions [Abramowitz and Stegun (1964)] for large arguments $b = \beta\kappa/a$

$$i_l(b) \sim \frac{1}{2b}e^b \left\{ 1 - \frac{l(l+1)}{2b} + \mathcal{O}(1/b^2) \right\}$$

results in

$$\hat{i}_l(b) \sim 1 - \frac{l(l+1)}{2b}$$

and generates a contribution of order a from the prefactor.

Collecting terms results in

$$\mathbf{U} = \mathbf{I} - \frac{1}{N}\frac{L_0}{\beta\kappa}\mathbf{J} = \mathbf{I} - \frac{1}{N}\frac{L_0}{\lambda}\mathbf{J},$$

where $\mathbf{I}$ is the unit matrix,

$$J_{ll'} = \frac{l(l+1)}{2}\delta_{ll'} - \beta\lambda f\frac{l'\delta_{l',l+1} + l\delta_{l',l-1}}{[(2l+1)(2l'+1)]^{1/2}}. \tag{2.9}$$

and I have made use of (1.13).

Using the last transformation allows me to write the matrix identity

$$\lim_{N\to\infty}\mathbf{U}^N = \lim_{N\to\infty}\left(1 - \frac{1}{N}\frac{L_0}{\lambda}\mathbf{J}\right)^N = e^{-L_0/\lambda\mathbf{J}}$$

and thus take the continuum limit.

It is now clear that the properties of the $\mathbf{J}$ matrix control the behavior of a the wormlike chain under the influence of an external force. A number of these properties are crucial:

- $\mathbf{J}$ is, like $\mathbf{U}$, real and symmetric;
- it follows that it has a nondegenerate spectrum of real eigenvalues

$$\Lambda_0 < \Lambda_1 \cdots < \Lambda_\nu \cdots$$

  with corresponding normalized eigenvectors $\{|\nu>\}$,
- and that it can be represented as

$$\mathbf{J} = \sum_\nu \Lambda_\nu |\nu><\nu| \quad ;$$

- consequently,

$$\hat{Z}_\infty(f) = \lim_{N\to\infty}\hat{Z}_N(f) = \sum_\nu e^{-L_0/\lambda\Lambda_\nu}\left|A_\nu^0\right|^2, \tag{2.10}$$

  where $A_\nu^0 =<\nu|0>$ is the $l = 0$ component of the νth eigenvector.

Two dimensionless quantities control thermodynamic behavior. First, the external force, which enters $\mathbf{J}$ in the dimensionless combination $\bar{f} = \beta\lambda f$, i.e. the ratio of elastic to thermal energy, and controls the spectrum of $\mathbf{J}$. Second, the size parameter L_0/λ which measures the length of the chain in units of the persistence length. If the chain is sufficiently large compared to the persistence length, the sum (2.10) will be dominated by the smallest eigenvalue Λ_0 [1] and — up to small, size-independent corrections —

$$\ln \hat{Z}_\infty(f) \approx -\frac{L_0}{\lambda}\Lambda_0 \qquad (2.11)$$

or, using, (2.4)

$$L = -\frac{L_0}{\beta\lambda}\frac{\partial\Lambda_0}{\partial f}, \qquad (2.12)$$

which can be recast in the dimensionless form

$$\frac{L}{L_0} = -\frac{\partial\Lambda_0}{\partial\bar{f}}. \qquad (2.13)$$

As it turns out, diagonalizing the matrix (2.9) is equivalent to a well-known quantum mechanical problem. I will proceed to exploit this equivalence.

2.3.2   The analogy with the quantum rotator

The quantum Hamiltonian of a rigid, isotropic free rotator can be represented in polar coordinates (θ, ϕ) by

$$\mathcal{H} \equiv \frac{\hbar^2}{2I}\mathcal{L}^2 = -\frac{\hbar^2}{2I}\left\{\frac{1}{\sin\theta}\frac{\partial}{\partial\theta}\left(\sin\theta\frac{\partial}{\partial\theta}\right) + \frac{1}{\sin^2\theta}\frac{\partial^2}{\partial\phi^2}\right\}$$

where $\hbar = h/(2\pi)$, h is the Planck constant, I the moment of inertia of the rotator and $\mathcal{L}$ the angular momentum operator.

Now suppose the quantum rotator carries an electric dipole moment $\vec{p}$. In the presence of external electric field $\mathcal{E}$ acting in the direction of the polar axis, the Hamiltonian will be modified by a term $-p\mathcal{E}\cos\theta^2$. Consequently, the Schrödinger equation will be of the form

$$\left[-\frac{1}{2}\frac{d}{dx}\left\{(1-x^2)\frac{d}{dx}\right\} - \bar{f}x\right]\psi(x) = \Lambda\psi(x) \qquad (2.14)$$

where I have (i) made the change of variables $x = \cos\theta$, (ii) rescaled the energies by a factor $\hbar^2/I$, (iii) introduced the dimensionless field strength

---

[1]If the condition $L_0 \gg \lambda$ does not hold, as is e.g. the case for stiff chains, the contributions arising from smaller eigenvalues must be explicitly taken into account. I note, parenthetically, here that the exact, full theory of the continuum WLC chain is an old, challenging mathematical problem which goes somewhat beyond the scope of this book; a central result is that the end-to-end distance distribution function and the scattering function (cf. 4.4) can be represented in terms of continuum fractions [Stepanow and Schütz (2002); Spakowitz and Wang (2004)]. From a practical point of view however, the eigenvector representation used in this book allows the accurate numerical computation of WLC distribution and scattering functions using standard mathematical subroutines.

[2]The model can be recognized as a rudimentary description of the Stark effect in a rigid diatomic molecule [von Meyenn (1970)]

$\bar{f} = Ip\mathcal{E}/\hbar^2$ and (iv) dropped the azimuthal dependence. The latter step restricts the eigenvalue spectrum to the $m = 0$ azimuthal quantum number sector - which is not a severe restriction if we are only looking for the lowest quantum state of the rotator.

We now immediately recognize (2.9) as the matrix representation of the dimensionless rotator Hamiltonian (2.14) in the space of Legendre polynomials $\{P_l(x)\}$.

2.3.2.1   *Limiting behavior at low fields*

The lowest order correction to the eigenvalue spectrum (quadratic, since the linear term vanishes on symmetry grounds) can be obtained by second order perturbation theory. In the limit $\bar{f} << 1$ the ground state energy behaves as

$$\Lambda_0(\bar{f}) \approx -\frac{1}{3}\bar{f}^2 \quad . \tag{2.15}$$

Inserting (2.15) in (2.13) leads to the linear, low-force limit of the force-extension relationship

$$\frac{L}{L_0} \approx \frac{2}{3}\frac{\lambda}{k_B T}f \quad . \tag{2.16}$$

2.3.2.2   *Limiting behavior at high fields*

In the limit $\bar{f} \to \infty$ the rotator is "hindered". Its motion consists of small oscillations around the field axis. For angles $\theta << 1$, (2.14) transforms to

$$-\frac{1}{2\theta}\frac{d}{d\theta}\left(\theta\frac{d\psi}{d\theta}\right) - \bar{f}\left(1 - \frac{1}{2}\theta^2\right)\psi = \Lambda\psi \tag{2.17}$$

which we recognize as formally identical to the *radial* part of the isotropic two-dimensional harmonic oscillator. The lowest eigenstate

$$\psi_0(\theta) \propto e^{-\bar{f}^{1/2}\theta^2/2}$$

and the corresponding eigenvalue

$$\Lambda_0(\bar{f}) \sim -\bar{f} + \bar{f}^{1/2} \tag{2.18}$$

can be found by simple inspection.

Inserting (2.18) in (2.13) leads to the saturation, high-force limit of the force-extension relationship

$$\frac{L}{L_0} \approx 1 - \frac{1}{2}\left(\frac{k_B T}{\lambda f}\right)^{1/2} \quad . \tag{2.19}$$

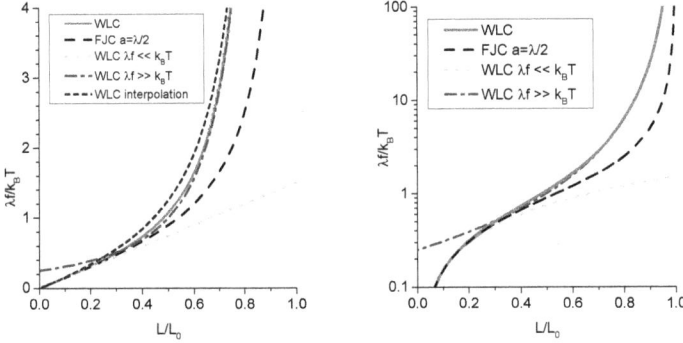

Fig. 2.1 *Left panel:* Continuous curve: the force-extension curve for the continuum limit of the Kratky–Porod chain, known as the wormlike chain (WLC). For comparison, the dashed curve represents the force extension curve, (2.8), for an "equivalent" FJC with segment length $a = 2\lambda$. Note that both curves have a common slope at the linear, low-field limit, and a common high field asymptote $L/L_0 \to 1$. Their behavior at intermediate and high fields differs significantly however. Also shown are the low- (dotted line (2.16)) and high-(dash-dotted line, (2.18)) force limits of the WLC model, as well as the interpolation formula (short-dashed line, (2.20)). *Right panel:* As in left panel, in semilog scale.

2.3.2.3 *Arbitrary field strength*

In the case of intermediate fields it is necessary to resort to a full numerical diagonalization of the matrix (2.9). The resulting force-extension curve is depicted in Fig. 2.1, along with low- and high-field asymptotics.

A useful interpolation between (2.16) and (2.19), which incorporates both limits and provides a good approximation for intermediate fields is provided by the formula [Bustamante *et al.* (1994)]

$$\frac{\lambda f}{k_B T} = \frac{1}{4}\frac{1}{(1 - L/L_0)^2} - \frac{1}{4} + \frac{L}{L_0}. \tag{2.20}$$

2.4 The force-extension relationship of the discrete KP model

In cases where the local bending stiffness is low, and the persistence length becomes comparable to — or lower than — the distance between successive monomers, the continuum approximation breaks down. In such cases it is not legitimate to use the WLC model; one must go back to the original KP formulation and derive the force-extension relationship in terms of the original KP model.

Going back to end of section 2.1, I expand, in analogy with (1.37) , the exponential

$$e^{\beta a f x} = \sum_{k=0}^{\infty} (2k+1) i_k (\beta a f) P_k(x), \quad |x| \leq 1, \tag{2.21}$$

in terms of modified spherical Bessel functions and Legendre polynomials. This leads, again in analogy with section 1.3, to a closed form expression for the matrix elements

$$U_{ll'} = \frac{1}{2} [\hat{i}_l(b)] \hat{i}_{l'}(b) (2l+1)(2l'+1)]^{1/2} u_{ll'},$$

$$u_{ll'}(\tilde{f}) = \sum_{k=|l-l'|, k+l+l'=2r}^{l+l'} (2k+1)(-1)^k \frac{1}{r+1/2}$$

$$\cdot \frac{\Psi(r-k)\Psi(r-l)\Psi(r-l')}{\Psi(r)} i_k(\tilde{f}), \tag{2.22}$$

where $\tilde{f} = \beta a f$. Now, if the number of monomers is sufficiently large, the partition function (2.6) will be dominated by the largest eigenvalue μ_0 of the real, symmetric matrix $\mathbf{U}$. The free energy will be

$$G = -N k_B T \ln \mu_0, \tag{2.23}$$

resulting in a force-extension relationship of the form

$$\frac{\Delta L}{L_0} = \frac{\partial \ln \mu_0}{\partial \tilde{f}}. \tag{2.24}$$

Note that the dimensionless force variables are defined differently in the KP, WLC, and FJC cases. This is because in each case the relevant length scale differs. In the first case it is the monomer distance, in the second the persistence length, and in the third the Kuhn length.

Numerical computation of the eigenvalue spectrum of $\mathbf{U}$ is straightforward and fast. Indeed, for most practical applications one may restrict the size of the matrix to less than 20×20.

2.5   The DNA force-extension relationship

2.5.1   *Double-stranded DNA*

One of the most exciting "single molecule" experiments of the 1990's was the direct mechanical measurement of the response of the double helix to an externally applied force [Smith *et al.* (1992)]. The measurement was made possible by attaching one end of a single DNA molecule (a dimer of the λ-phage, composed of $2 \times 48.5 = 97$ thousands of base pairs) to a glass surface and the other end to a magnetic bead. Owing to the stiffness of ds-DNA, a force of only 0.1 pN was sufficient to stretch the molecule to about half its contour length.

At force levels between 0.03 and 20 pN, the elastic behavior was found to follow closely the WLC model[3] (cf. Fig. 2.2) with an observed persistence length of 53.4 nm and a contour length $3.38\text{Å} \times 97000 = 32.8\,\mu\text{m}$, consistent with crystallographic data[Bustamante *et al.* (1994)].

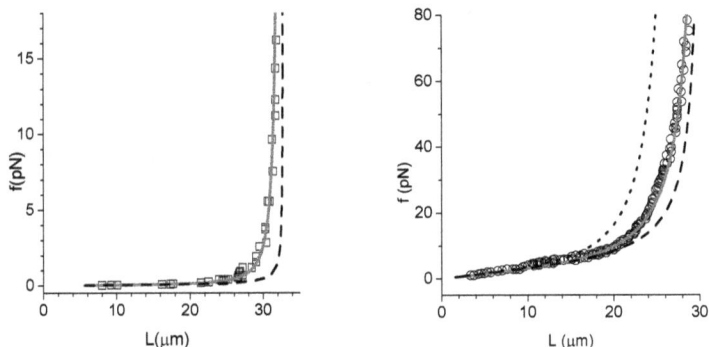

Fig. 2.2 *Left panel:* The force-extension curve for ds-DNA. The squares are data points from [Bustamante *et al.* (1994)] and represent measurements for the 97 kb long λ-phage dimer. The solid curve is a numerically computed WLC force-extension curve, (2.13), at $T = 20C$, with a monomer length 3.38 Å and a persistence length 53.4Å. The dashed curve describes an equivalent FJC model (2.8) with a segment length 106.8Å and the same contour length. *Right panel:* The force-extension curve for ss-DNA. The squares are data points from [Smith *et al.* (1996)] and represent measurements for the 48.5 kb long λ-phage. The solid curve is a numerically computed KP force-extension curve (2.24) with a monomer length 6.3 Å, as estimated from crystallographic data on ss-DNA [Murphy *et al.* (2004)], and a persistence length 6.7Å. The dashed curve describes an equivalent FJC model (2.8) with a segment length 1.34 nm and the same contour length. The dotted curve is a numerically computed WLC curve with the same parameters as the KP curve.

---

[3]In all physical examples considered within the scope of this book, the elastic behavior of DNA will be adequately characterized by its *bending* and, sometimes, its *stretching* stiffness. This is clearly an oversimplification. *Twisting* rigidity is an important feature of the elastic response of the double helix and, is almost always -except at the lowest applied force levels- coupled to both stretching and bending [Marko (1997); Nomidis *et al.* (2017)].

2.5.2 Single-stranded DNA

Single-stranded DNA (ss-DNA) offers far less resistance to bending than ds-DNA. Estimates of its persistence length vary, depending on salt concentration between 1 and 3 nm. Its force-extension curve is shown in Fig. 2.2 (right panel). The data from a single λ-phage molecule (48.5 kb, Na^+ concentration 0.150 M) [Smith *et al.* (1996)] follow quite closely the discrete KP numerical calculation (2.24) performed with a monomer distance $a = 0.63$ nm, as estimated from crystallographic data on ss-DNA [Murphy *et al.* (2004)], and a stiffness constant $J = 5.259$ pN nm, corresponding to a room temperature persistence length $\lambda = 0.67$ *nm* (cf. (1.12)). The figure also shows, for comparison, the equivalent FJC curve, computed with a segment length 2λ, and the numerically calculated WLC continuum limit curve. It should be pointed out that, because in this case the persistence length is comparable to the monomer distance, the continuum approximation leading to the WLC limit is not valid; therefore the KP and WLC results differ quite substantially.

Chapter 3

DNA packaging and wrapping

3.1 Packaging of genomic material in a DNA virus

Viruses are simple biological entities carrying a short genome (RNA or DNA, of the order of ten genes) encoding their constituent proteins. Their ubiquitous association with disease masks an enormous contribution to the creation and exchange of genetic information[Nasir *et al.* (2014)]. Viruses that carry double-stranded DNA as their genomic material - known as DNA viruses - have some striking structural and mechanical properties. Typically, a DNA chain consisting of tens of thousands of base pairs (about 10 microns in stretched length) has to be packaged in a capsid with diameter of the order of the persistence length. This very close, comparable to crystalline, packing results in high bending energies and pressures at the capsid walls. Accordingly, high forces of the order of 50 pN are necessary for packaging, and therefore, powerful molecular motors must be at work. In what follows I will describe in some detail a model calculation that describes the energetic balance in the ϕ29 bacteriophage. The calculation follows, with some simplifications, the work of Purohit et al [Purohit *et al.* (2003)].

3.1.1 *Shape and parameters*

The bacteriophage ϕ29 is a ds-DNA virus consisting of 1.93×10^4 base pairs. If stretched it would have a length $L_0 = 6.58\mu$m. The capsid is prolate in shape with an outer diameter of about 42 nm and height 54 nm [Tao *et al.* (1998)]. Following [Purohit *et al.* (2003)] I will approximate its interior by a cylinder of radius $R_{out} = 19.4$nm and height $h = 37.9$nm, thereby roughly preserving the capsid volume. The configuration of the packaged DNA will be assumed to be a reverse spool with concentric rings of decreasing radii, starting from the capsid boundaries and proceeding towards the center (cf.

Fig. 3.1). As the chain is spooled, successive segments of the chain are oriented parallel to each other in a hexagonal close packing order, with a vertical spacing d_s and a horizontal spacing $\sqrt{3}/2d_s$. Because of the cylindrical shape, if the packing has reached the radius R, there will be $\nu(R') = h/d_s$ rings for any $R' > R$ and zero rings at larger $R' < R$. The length of the packaged virus will be related to R via

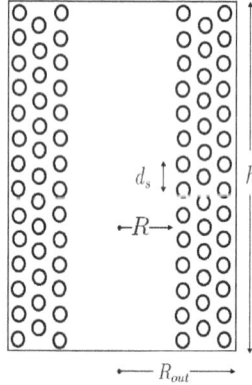

Fig. 3.1   Simplified view of a cross-sectional cut of a viral capsid (after [Purohit *et al.* (2003)]). The circles represent strands of DNA that point into and emerge out of the page. R_{out} is the outer radius of the cylindrical capsid, h its height and d_s the vertical spacing between the strands.

$$L = 2\pi \sum_i R_i \nu(R_i) \approx 2\pi \frac{2}{\sqrt{3}d_s} \int_R^{R_{out}} dR' R' \frac{h}{d_s} = \frac{2\pi h}{\sqrt{3}d_s^2}(R_{out}^2 - R^2),$$

(3.1)

where I have approximated the sum by an integral. Note that I can invert (3.1) to obtain R as a function of the packaged length L:

$$\left(\frac{R}{R_{out}}\right)^2 = 1 - \frac{\sqrt{3}d_s^2}{2\pi h R_{out}^2}L.$$

(3.2)

One more comment is in order here. So far I have simply described a geometry of packaging. The actual value of the stacking parameter d_s remains to be determined, and this can only be done by taking into account the energies involved. However, one constraint on the allowed values of d_s comes from geometry and is already evident. If the right hand side of (3.2) is to remain positive then at complete packaging, i.e. $L = L_0$, the inequality

$$d_s < \left(\frac{2\pi h}{\sqrt{3}L_0}\right)^{1/2} R_{out} \equiv d_0 = 2.804\, nm$$

(3.3)

must hold. This exceeds the diameter of ds-DNA, 2.1 nm, only by about one third; in other words, packing will be very close indeed.

3.1.2 Bending vs. hydration energies. Packaging force. Pressure on capsid walls

3.1.2.1 Bending energy of an elastic ring

Consider an element of length ds of a WLC with a local curvature $1/R$. Its bending energy will be, according to (1.4),

$$\frac{1}{2}\kappa\frac{1}{R^2}ds.$$

If the curvature is constant, as in a ring of radius R, the total bending energy of the ring will be equal to

$$\pi k_B T\frac{\lambda}{R}.$$

3.1.2.2 Bending energy of the packaged length

Summing over all rings of the packaged length gives the total elastic (bending) energy of the packaged length

$$E_{el} = \pi\lambda k_B T\sum_i\frac{1}{R_i}\nu(R_i) \approx \pi\lambda k_B T\frac{2}{\sqrt{3}d_s}\int_R^{R_{out}}dR'\,\frac{1}{R'}\frac{h}{d_s}$$

$$= \frac{2\pi\lambda h}{\sqrt{3}d_s^2}k_B T\ln\left(\frac{R_{out}}{R}\right). \tag{3.4}$$

Alternatively, making use of (3.2), it is possible to express the total elastic energy in terms of the packaged length L:

$$E_{el}(L,d_s) = -\frac{\pi\lambda h}{\sqrt{3}d_s^2}k_B T\ln\left(1 - \frac{\sqrt{3}d_s^2}{2\pi h R_{out}^2}L\right). \tag{3.5}$$

3.1.2.3 Hydration forces

Neighboring segments of the spooled ds-DNA in solution are not independent of each other. Like other molecular or macromolecular charged solutes DNA forms an aqueous layer which surrounds the double helix, and screens the excess charges. As a result, two neighboring chain segments placed parallel to each other at a distance d_s will repel each other [Leikin *et al.* (1993)] with a force per unit length approximately given by

$$f_0\,e^{-d_s/b}$$

for separations d_s much larger than the screening length b. The interaction energy (per unit length) of pair of two segments can then be computed from the work necessary to bring them together

$$-f_0 \int_\infty^{d_s} dx' \, e^{-x'/b} = f_0 b e^{-d_s/b}.$$

For a ring of radius R in the spool this leads to a hydration interaction energy

$$2\pi R \times 6 \times \frac{1}{2} \cdot f_0 b e^{-d_s/b},$$

where the factor 6 comes from the hexagonal close packing (i.e. each ring interacts with six neighbors at a distance d_s), and the $1/2$ is to avoid double counting.

3.1.2.4 *Hydration energy of the packaged length*

Summing over all rings, in analogy with (3.4), I obtain the hydration energy for the total packaged length

$$E_{hy} = 6\pi f_0 b e^{-d_s/b} \sum_i R_i \nu(R_i) \approx 6\pi f_0 b e^{-d_s/b} \frac{2}{\sqrt{3} d_s} \int_R^{R_{out}} dR' R' \frac{h}{d_s}$$

$$= \frac{2\sqrt{3}\pi h}{d_s^2} \left(R_{out}^2 - R^2 \right) f_0 b e^{-d_s/b}. \quad (3.6)$$

Alternatively, making use of (3.2), it is possible to express the total energy of hydration in terms of the packaged length L:

$$E_{hy}(L, d_s) = 3 f_0 b L e^{-d_s/b}. \quad (3.7)$$

Note that I could have written down Eq. (3.7) directly by observing that the whole length of the spool has a hydration energy arising from its local hexagonal coordination.

Following [Leikin *et al.* (1993); Purohit *et al.* (2003)] I will use $f_0 = 1.25 \times 10^5 \, pN/nm$ and $b = 0.27 \, nm$. These values should be representative of DNA solutions at a Na^+ molar concentration of .5 M and have been determined from osmotic pressure measurements.

3.1.2.5 *Hydration vs. bending. Packaging force*

The optimal spacing of the spool comes as a result of minimizing the sum of bending and hydration energies. The minimization condition

$$\frac{\partial}{\partial d_s} \left[E_{el}(L, d_s) + E_{hy}(L, d_s) \right] = 0$$

or, after some rewriting,

$$3\frac{f_0 R_{out}^2}{\lambda k_B T} d_s e^{-d_s/b} = \left(\frac{d_0}{d_s}\right)^2 \frac{L_0}{L} \cdot \ln\left[1 - \left(\frac{d_s}{d_0}\right)^2 \frac{L}{L_0}\right] + \frac{1}{1 - \left(\frac{d_s}{d_0}\right)^2 \frac{L}{L_0}} \quad (3.8)$$

determines the equilibrium spacing d_s as a function of packaging length L. The result is shown in Fig. 3.2, Note that the limiting value of the equilibrium spacing d_s at complete packing is equal to 2.796 nm, barely above the limiting value d_0. This corresponds closely to the observed value of 2.75 nm obtained from structural data [Earnshaw and Casjens (1980)].

It is now possible to determine the packaging force as the derivative of the total energy with respect to the packaged length. Note that this only involves the explicit dependence on L, i.e.

$$F = \frac{d}{dL}E = \left(\frac{\partial E}{\partial L}\right)_{d_s} + \left(\frac{\partial E}{\partial d_s}\right)_L \cdot \frac{dd_s}{dL} = \left(\frac{\partial E}{\partial L}\right)_{d_s}$$

where the second term vanishes identically due to the minimization condition. The force is then given by

$$F = \frac{\lambda k_B T}{2R_{out}^2} \frac{1}{1 - \left(\frac{d_s}{d_0}\right)^2 \frac{L}{L_0}} + 3f_0 b e^{-d_s/b} \quad (3.9)$$

where d_s is understood to be given by the condition (3.8).

The packaging force as a function of the packaged length is also shown in Fig. 3.2. It reaches a maximum of $58.4 pN$ at complete packing. Again, this corresponds closely to the value observed in single-molecule experiments [Smith *et al.* (2001)]. It is also worth noting that this is one of the strongest forces observed in natural molecular motors.

3.1.2.6 *Energies*

It is possible to compute the value of the area under the force-packaged length curve of Fig. 3.2 and get an idea of the energy stored in the packaged virus. The same can be done by computing separately bending and hydration energies (at the equilibrium spacing d_s) with the added advantage that we gain insight at the relative importance of bending and hydration energies at all packing fractions. The result is shown in Fig. 3.3. At low packing fractions, the dominant contribution to the energy comes from bending. Hydration energies become comparable to bending energies around $L/L_0 = 0.6$. Thereafter, hydration energies dominate the picture.

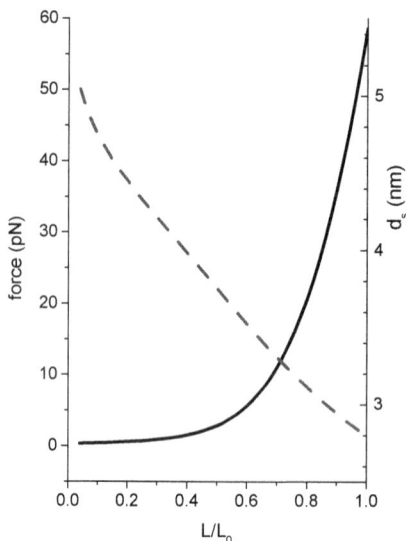

Fig. 3.2   The packaging force (continuous curve, left y-scale) as a function of packed fraction. Note the very gradual increase at low packing fraction. Also shown is the equilibrium spacing (dashed curve, right y-scale) as a function of packed fraction.

3.1.2.7   *Pressure on the capsid walls*

It is possible to get an estimate of the forces exerted on the capsid walls by looking at the dependence of the total energy on the capsid dimensions at constant packing fraction. Consider for example a vertical slice of the capsid, of angular width $\Delta\theta$, volume $\Delta V = 1/2R_{out}^2 h \Delta\theta$ and surface area $\Delta S = h R_{out} \Delta\theta$. Assuming that the packaged virus is homogeneous in the direction of the cylindrical axis, the energy ΔE stored in this slice will be a fraction $\Delta V/(\pi R_{out}^2 h) = \Delta\theta/(2\pi)$ of the total energy. The force exerted on the wall element in the radial direction by the packaged virus can be calculated as the derivative

$$\Delta F_{radial} = - \frac{d\Delta E}{dR_{out}}$$

$$= \frac{\lambda L}{2\pi R_{out}^2} \frac{1}{1 - \left(\frac{d_s}{d_0}\right)^2 \frac{L}{L_0}} k_B T \Delta\theta,$$

corresponding to a pressure on the vertical capside wall

$$p = \frac{\Delta F_{radial}}{\Delta S} = \frac{\lambda L}{2\pi R_{out}^4 h} \frac{1}{1 - \left(\frac{d_s}{d_0}\right)^2 \frac{L}{L_0}} k_B T. \tag{3.10}$$

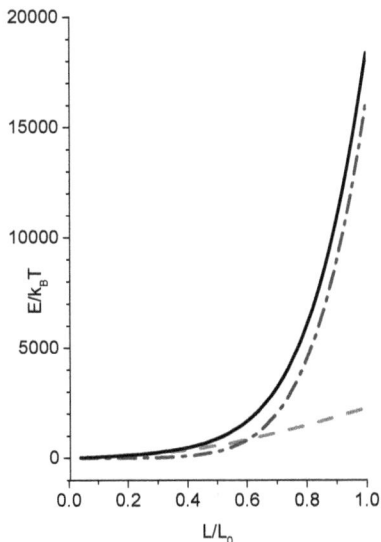

Fig. 3.3 Energies involved in the packaging process, expressed as multiples of $k_B T$, as a function of packed fraction. The total energy (solid curve) consists of bending (dashed line) and hydration (dashed-dotted line) contributions. The total energy at full packing is about 18300 k_BT.

Note that the dependence of d_s on R_{out} does not enter the calculation, since d_s is such as to satisfy the minimization condition. Therefore, since the hydration energy has no explicit dependence on R_{out}, only the bending energy contributes. This does not however mean that hydration energies are irrelevant! They determine the value of d_s which enters (3.10).

At full packing (3.10) evaluates to 70 at. This extremely high pressure, in agreement with other estimates and measurements in the order of tens of atmospheres, imposes significant material strength requirements on the capsid membrane. Moreover, it appears to be instrumental in the process of ejecting viral DNA by literally blasting it into the cell and bringing about infection [Bauer *et al.* (2013)].

3.2 Wrapping in nucleosomes

3.2.1 *Genomic packaging*

At the higher end of the evolutionary scale things get even more crowded. If the human genome, i.e. a stretched DNA length of the order of two meters, is going to fit into the nucleus of each cell, with a size of the order of ten microns, while still available for selected functional tasks, a hierarchical packing order spanning five orders of magnitude will be necessary. This packing is accomplished in eukaryotic cells by a protein-DNA complex called chromatin. Chromatin is composed of fundamental repeating units called nucleosomes. Nucleosomes are the lowest ranking members of

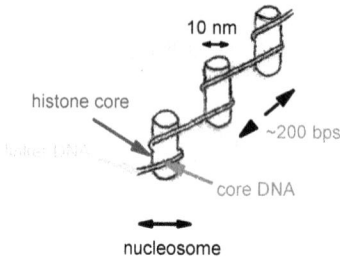

10 nm

histone core

~200 bps

core DNA

nucleosome

Fig. 3.4 Schematic representation of DNA compaction at the $10\,nm$ fiber level ("beads on a string"). The nucleosome core consists of the histone octamer, along with the superhelical DNA. It has a roughly cylindrical shape with a height of $6\,nm$ and a diameter of $11\,nm$. Successive nucleosomes are joined together by segments of linker DNA, consisting of 20-80 base pairs.

the packing hierarchy. They consist of a histone octamer around which an approximately 146 base pairs long segment of the double helix is "superhelically" wrapped 1.67 times. Adhesion of the wrapped DNA to the octamer occurs at 14 sticking points, essentially following the pitch of the double helix. The diameter of the superhelix, as determined by X-ray crystallography, is $8.4\,nm$ [Richmond and Davey (2003)]. The histone octamer is composed of two copies of four histone proteins. Successive nucleosomes are joined together by segments of linker DNA, typically of length between 20 and 80 base pairs. The nucleosome core, i.e. the histone octamer (excluding the tails) along with the superhelical DNA (i.e. the 146 base pairs wound around the octamer) has a roughly cylindrical shape with a height of $6\,nm$ and a diameter of $11\,nm$. The resulting picture at this level of compaction, schematically depicted in Fig. 3.4, is that of the "$10\,nm$ fiber", also known as "beads on a string". The packing ratio achieved can be estimated by comparing the size of the bead (diameter of the nucleosome core), $11\,nm$, to the $68\,nm$ length of a $200-$base-pair long segment of the double helix, to be approximately 6.

The next level in the hierarchical packing structure of chromatin is achieved under physiological conditions, i.e. with increasing salt concentration. The "beads on a string" fiber swells to what appears to be a solenoidal structure with about 6 nucleosomes per turn. This is known as the 30 nm fiber. Further compactification involves extensive looping and scaffolding of the fiber, up to the scale of the chromosome.

3.2.2 Bending energies at the nucleosome core

Even a cursory look at the nucleosome size suggests that the wrapping of DNA in a cylinder of 11 nm diameter, i.e. five times smaller than the persistence length, must involve very strong bending. The bending energy can be estimated from (cf. 3.1.2)

$$E_{bending} = \frac{L\lambda}{2R^2} k_B T \qquad (3.11)$$

where L is the length of wrapped DNA, $R = 4.2$ nm and $\lambda = 50$ nm. According to the X-ray crystallographic evidence [Richmond and Davey (2003)] the wrapped part of DNA corresponds to 126 base pairs, because the last 10 base pairs on each side are essentially straight. This gives an $L = 126 \times 0.34$ nm $= 42.8$ nm and results in a bending energy equal to 60 $k_B T$.

Although the binding energies of the DNA to the histone core are not known exactly, it is clear that they must somehow compensate for this large amount of elastic energy, and provide a surplus for stability. The surplus need not be large, because the nucleosome must be a stable but dynamic structure, i.e. relatively open for functional tasks.

3.2.3 Unwrapping the nucleosome

Operating at the single molecule level, it has been possible[Mihardja *et al.* (2006)] to literally unwrap the nucleosome by applying an external force. Unwrapping proceeds in two steps. The first step, at a force of 3 pN, is reversible and corresponds to a free energy change of 24.7 kJ/mol or approximately 10 $k_B T$. The second step occurs at higher forces with a broad distribution centered at $8-9 pN$ and is not reversible. The interpretation of the experimental findings is that the first step corresponds to the unwrapping of the outer part of the DNA, whereas the second step unwraps the remaining part. If the first part unwraps approximately 1/3 of the DNA (half a turn in a total of 1.5 turns), this suggests a net free energy change of -30 $k_B T$ for nucleosome dissociation.

It is now possible to estimate how strong the adhesion forces holding the DNA to the octamer must be. The 14 sticking points must supply a total binding energy of 30 $k_B T$ in addition to compensating about 60 $k_B T$ of bending energy. This corresponds to approximately 6 $k_B T$ per sticking point.

Chapter 4

Scattering from DNA in solution

4.1 Elastic scattering from dilute solutions of macro-molecules

Generally, in a scattering experiment, one measures the differences in energy and direction between incident and scattered radiation beam owing to the presence of a sample. If scattering is elastic there is no detectable change in energy between incident and scattered radiation. The wavevector of incident radiation $\vec{k}_i$ has the same magnitude as that of the scattered radiation, $\vec{k}_f$,

$$|\vec{k}_f| = |\vec{k}_i| = k.$$

The scattering wavevector is defined as the difference $\vec{q} = \vec{k}_f - \vec{k}_i$. The three vectors can be viewed as forming an isosceles triangle with an apex angle θ, the scattering angle. The scattering geometry, depicted schematically in Fig. 4.1, is reflected in the relationship

$$q \equiv |\vec{q}| = k \sin(\theta/2).$$

An elastic scattering experiment consists of measuring the scattered intensity at a certain direction, $I(\vec{q})$, as a function of the scattering wavevector, or, equivalently, scattering angle. By definition, it does not probe the dynamics of the sample. However, it can be used to yield structural information at the scale of the radiation wavelength used. In a dilute solution of identical macromolecules, i.e. where interference from scattering originating in different molecules can be neglected, scattering is incoherent. One observes the sum of scattering intensities of each molecule. The angular distribution of the scattering intensity therefore reflects the average

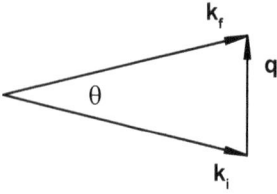

Fig. 4.1   The geometry of an elastic scattering experiment. The wavevectors of the incident and scattered beams, $\vec{k}_i$ and $\vec{k}_f$, respectively, form the two legs of an isosceles triangle. The angle θ between them is the scattering angle. The difference $\vec{q}$ is the scattering wavevector.

intramolecular structure. This is expressed by the relationship

$$\frac{I(\vec{q})}{I(0)} = S(\vec{q}) \equiv \frac{1}{(N+1)^2} \sum_{i,j=0}^{N} < e^{i\vec{q}\cdot(\vec{R}_i - \vec{R}_j)} >$$

$$= \frac{1}{N+1} + \frac{2}{(N+1)^2} \sum_{j=2}^{N} \sum_{i=0}^{j-1} < e^{i\vec{q}\cdot(\vec{R}_i - \vec{R}_j)} >, \qquad (4.1)$$

where N is the number of monomers in each macromolecule, the double summation runs over all monomers, $\vec{R}_i$ denotes the position of the ith monomer, the angular brackets denote an ensemble average, $I(0)$ is the forward scattering intensity and $S(\vec{q})$ the structure factor of the macromolecule.

4.2   The structure factor of the Kratky–Porod chain

4.2.1   *An intermediate result*

Consider a typical term $< e^{i\vec{q}\cdot(\vec{R}_i - \vec{R}_j)} >$ in the sum (4.1) with $i < j$. In order to evaluate the average required, we need the probability distribution of the vector $\vec{R}_j - \vec{R}_i$. This appears to be a somewhat more general case of the end-to-end distance probability distribution $P_N(r)$ or, more precisely, $P_N(r|\vec{R}_N - \vec{R}_0)$, described at length in section 1.3. However, it is straightforward to observe that the integrals over the directions of segments with index smaller than i or larger than j, i.e. the outer parts of the chain, factor out in exactly the same way in both numerator and denominator of the generalized version of (1.28). Only "internal" variables matter. In other words, what we need is not $P_N(r|\vec{R}_j - \vec{R}_i)$, but $P_{j-i}(r|\vec{R}_j - \vec{R}_i)$, an "end-to-end" probability distribution defined in a reduced chain with $j - i$ segments. In terms of this probability distribution

$$
\begin{aligned}
< e^{i\vec{q}\cdot(\vec{R}_i - \vec{R}_j)} > &= \int d\vec{r}\, P_{j-i}\left(r|\vec{R}_j - \vec{R}_i\right) e^{i\vec{q}\cdot\vec{r}} \\
&= P_{j-i}(\vec{q}|\vec{R}_j - \vec{R}_i) \\
&= (\mathbf{F}^{j-i})_{00},
\end{aligned}
\tag{4.2}
$$

in terms of the $\vec{q}$-dependent matrix $\mathbf{F}$ which has been defined in 1.3.

4.2.2 *Structure factor of the homogeneous KP chain*

It is now straightforward, using (4.2), to obtain a closed-form expression for the structure factor of the KP chain in terms of $\mathbf{F}$. Noting that the power $\mathbf{F}^l$ occurs exactly $N - l + 1$ times in the double sum (4.1), we can rewrite

$$
S(\vec{q}) = \frac{1}{N+1} + \frac{2}{(N+1)^2} \sum_{l=1}^{N}(N - l + 1)(\mathbf{F}^l)_{00}.
\tag{4.3}
$$

Eq. 4.3 provides a basis for the numerical evaluation of the Kratky–Porod structure factor in the cases where the model's discrete nature is essential, i.e. when the distance a between successive monomers is comparable to the persistence length. If a continuum approximation is justified, the WLC limit can be taken. This will be described in section 4.4.

4.3 Structure factor of the freely jointed chain (FJC)

In the limiting case of the FJC, the matrix $\mathbf{F}$ reduces to the scalar $j_0(qa) = \sin qa/qa$. The sum (4.3) can be performed explicitly and leads to

$$
S(\vec{q}) = \frac{1}{N+1}\frac{1+j_0}{1-j_0} + \frac{2}{(N+1)^2}\frac{j_0}{(1-j_0)^2}(j_0^{N+1} - 1),
\tag{4.4}
$$

shown in the left panel of Fig. 4.2 for $N = 30$.

In the limit of long wavelengths, the scattering function samples only the effective size of the polymer chain, as described by the radius of gyration, i.e.

$$
\frac{I(q)}{I(0)} \equiv S(q) \approx e^{-q^2 R_g^2/3},
\tag{4.5}
$$

independently of whether a chain is locally stiff or freely jointed. This is a general feature of the Guinier limit, $qR_g \lesssim 1$. Consequently, if scattering data are available in this regime, a Guinier analysis, i.e. a plot of the logarithm of the scattering intensity vs. q^2, can be used to estimate R_g and $I(0)$.[1] The right panel of Fig. 4.2 exhibits the Guinier regime of the FJC structure factor (4.4), showing also, for comparison, the straight line $-\frac{1}{3}R_g^2 q^2$ with R_g from (1.24).

A useful approximation to (4.4) is the Debye function

$$S_{Debye}(q) = \frac{2}{(qR_g)^4}[e^{-(qR_g)^2} - 1 + (qR_g)^2], \qquad (4.6)$$

also shown in the left panel of Fig. 4.2. Note that the Debye function describes correctly the q^{-2} decay characteristic of random (Gaussian) behavior, while it fails to account for the oscillatory behavior of (4.4) at high wavevectors, or for its finite limiting value in the limit $q \to \infty$.

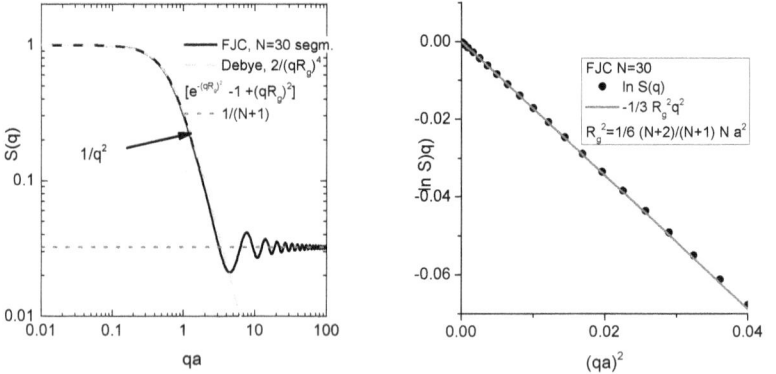

Fig. 4.2   *Left panel:* The structure factor of the FJC with $N = 30$ segments (continuous line). Also shown are the $q \to \infty$ limit (dotted line) and the Debye approximation, (4.6). *Right panel:* The Guinier regime of the FJC. The exact $S(q)$ function, (4.4) (circles) and the straight line $-\frac{1}{3}R_g^2 q^2$ with $R_g \approx 2.27$ from (1.24). Deviations from exact Guinier-like behavior become visible at the maximum q displayed, 0.2, where $qR_g \approx 0.45$.

---

[1]The forward scattering intensity is proportional to the total number of scatterers. Thus, for example, in the case of x-ray scattering from a macromolecule in solution, it can be estimated from the difference in electron density between solute and solvent, the concentration of the macromolecule and its molecular weight. Alternatively, if a macromolecule of known molecular weight is used for calibration purposes, the Guinier analysis may serve as a method of determining molecular weights.

4.4   Structure factor of the WLC

4.4.1   *The approach to the continuum limit*

In the limit of $N \gg 1$ it is possible to substitute the summation in (4.3) by an integral over the continuous variable $s = la = lL_0/N$. This transforms (4.3) to

$$S(\vec{q}) \approx \frac{2}{L_0} \int_0^{L_0} ds \left(1 - \frac{s}{L_0}\right) (\mathbf{F}^{Ns/L_0})_{00}.$$

In the continuum limit, $N \to \infty$, $a \to 0$, $Na \to L_0$, the above substitution becomes exact. Moreover, it is sufficient to keep only the lowest order term in an expansion of $\mathbf{F}$ to a power series in a. The argument proceeds exactly as in section 2.3.1 - with the substitution $\beta f \leftrightarrow -iq$ - and results, to leading order in a, in

$$\mathbf{F}(\vec{q}) = \mathbf{I} - \frac{1}{N}\frac{L_0}{\lambda}\hat{\mathbf{J}}$$

where $\mathbf{I}$ is the identity matrix,

$$\hat{J}_{ll'} = \frac{l(l+1)}{2}\delta_{ll'} - i\lambda q \frac{l'\delta_{l',l+1} + l\delta_{l',l-1}}{[(2l+1)(2l'+1)]^{1/2}}, \tag{4.7}$$

and

$$\lim_{N\to\infty} \mathbf{F}^N = \lim_{N\to\infty} \left(1 - \frac{1}{N}\frac{L_0}{\lambda}\hat{\mathbf{J}}\right)^N = e^{-L_0/\lambda\hat{\mathbf{J}}}. \tag{4.8}$$

The latter property allows me to rewrite the structure factor of the WLC as

$$S(\vec{q}) = \frac{2}{L_0} \int_0^{L_0} ds \left(1 - \frac{s}{L_0}\right) (e^{-s/\lambda\hat{\mathbf{J}}})_{00}. \tag{4.9}$$

In contrast to the real, symmetric matrix $\mathbf{J}$ of section 2.3.1, $\hat{\mathbf{J}}$ is a complex, symmetric, *non-Hermitian* matrix. Complex symmetric matrices can, in general, be diagonalized using standard numerical techniques; therefore, $\hat{\mathbf{J}}$ can be equally useful in the present context as $\mathbf{J}$ was in deriving the force-extension relation. In analogy with section 2.3.1, $\hat{\mathbf{J}}$ will have the following properties:

- Its spectrum is nondegenerate and consists, in general, of both real and complex eigenvalues $\{\Lambda_\nu\}$. Complex eigenvalues appear as pairs of complex conjugates.

- If $|\nu>$ is a right eigenvector

$$\hat{\mathbf{J}}|\nu>=\Lambda_\nu|\nu>,$$

its *transpose* $<\nu|$, with elements $<\nu|i>=<i|\nu>$, will be a left eigenvector

$$<\nu|\hat{\mathbf{J}}=<\nu|\Lambda_\nu.$$

- It has a representation

$$\hat{\mathbf{J}}=\sum_\nu \frac{\Lambda_\nu}{<\nu|\nu>}|\nu><\nu| \tag{4.10}$$

in terms of its eigenvalues and an associated set of orthogonal eigenvectors. Note however that the inner product $<\nu|\nu>$ is not a real and positive norm - as would have been the case for a Hermitian matrix. It is, in general, a complex number.

- As a consequence, any diagonal matrix element of a function of $\hat{\mathbf{J}}$ can be represented as

$$\left(\mathcal{F}(\hat{\mathbf{J}})\right)_{ii}=\sum_\nu \mathcal{F}(\Lambda_\nu)A_\nu^{(i)} \tag{4.11}$$

where

$$A_\nu^{(i)}=\frac{<\nu|i>^2}{<\nu|\nu>}=\frac{<\nu|i>^2}{\sum_j<\nu|j>^2}$$

and $<\nu|i>$ is the *ith* component of the νth eigenvector.

The latter property allows me to perform the integral in (4.9) and rewrite it as

$$S(\vec{q})=2\sum_\nu \left[\frac{\bar{\lambda}}{\Lambda_\nu}-\left(\frac{\bar{\lambda}}{\Lambda_\nu}\right)^2\left(1-e^{-\frac{\Lambda_\nu}{\bar{\lambda}}}\right)\right]A_\nu^{(0)}, \tag{4.12}$$

where $\bar{\lambda}=\lambda/L_0$ is the dimensionless ratio of persistence length and polymer contour length.

4.4.2 *Numerical evaluation of the structure factor*

The sum (4.12) can be performed numerically as follows. First, the infinite matrix $\hat{\mathbf{J}}$ must be truncated to finite dimension l_{max}. It is then possible to use standard numerical diagonalization routines to obtain eigenvalues and eigenvectors. For all practical purposes, an order of approximately $l_{max}=15$ is sufficient and the computation is very rapid. Fig. 4.3 shows some characteristic results of the numerical summation, which agree quite well with Monte–Carlo (MC) simulations[2].

It is sometimes convenient to represent scattering data graphically as *Kratky plots*, plotting the function $qS(q)$ vs. qL. In such a plot, all curves start off linearly at the origin with unit slope. If the data follow the WLC model, the curve asymptotically approaches π as $qL \to \infty$. The degree of stiffness determines the offset and extent of the transition region. An example is shown in Fig. 4.4.

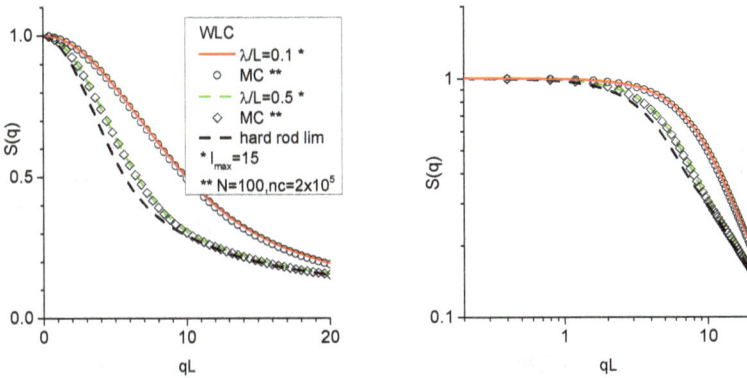

Fig. 4.3 *Left panel:* The structure factor of a flexible ($\lambda/L_0 = 0.1$) and a stiff ($\lambda/L_0 = 0.5$) WLC chain, computed from (4.12) via numerical diagonalization of the truncated matrix $\hat{\mathbf{J}}$ to a finite dimension $l_{max} = 15$. Also shown are results of numerical MC simulations of corresponding discretized versions of the WLC chains, using 100 segments and 2×10^5 configurations. The "hard rod" limiting case is plotted as a dashed line. *Right panel:* Same as left panel, in a double logarithmic scale.

4.5 Numerical evaluation of the end-to-end distance probability distribution function

The techniques developed in the previous section allow us to compute numerically the end-to-end distance probability distribution function of the WLC. Using (4.8) in (1.36) results in

$$P(\vec{q}) = \sum_{\nu} e^{-\frac{\Lambda_\nu}{\lambda}} A_\nu^{(0)}, \tag{4.13}$$

───────────

[2]A brief introduction to MC simulations of the WLC chain is provided in Appendix A.

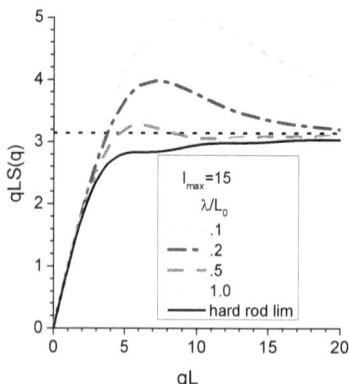

Fig. 4.4   An alternative plot of the WLC structure factor, using the product $qS(q)$ (Kratky plot), again for a variety of values of λ/L_0. In such a plot, all curves start off linearly at the origin with unit slope and asymptotically approach π as $qL \to \infty$. Also shown is the hard-rod limit (continuous line). The degree of stiffness determines the offset and extent of the transition region.

which can be then Fourier-transformed to yield the end-to-end probability distribution $P(r)$. Typical results of the numerical evaluation for a stiff chain, performed at varying levels of truncation of the infinite matrix $\hat{\mathbf{J}}$ are shown in Fig. 4.5. A truncation level $l_{max} = 15$ turns out to be sufficient for most practical purposes.[3]

4.6   Structure factors of simple geometrical molecules

The analysis of scattering experiments has been substantially facilitated by model calculations which refer to simple geometrical shapes, e.g. spheres or cylinders. If the density can be assumed to be uniform inside the molecule and zero outside it, the structure factor(4.1) reduces to

$$S(\vec{q}) = \left\langle \left| \frac{1}{V} \int_V d\vec{r}\, e^{i\vec{q}\cdot\vec{r}} \right|^2 \right\rangle \tag{4.14}$$

where the integral is taken over the molecular volume and the angular brackets denote an average over all possible orientations of the molecule.

---

[3]Note however that because of the oscillatory character of $P_N(q)$, some extra care must be taken with the numerical Fourier transform procedure. This results in somewhat higher computing times.

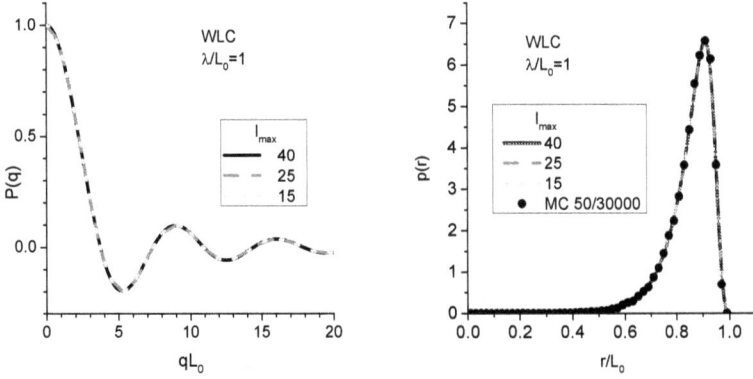

Fig. 4.5 *Left panel:* The Fourier transform $P(q)$ of the end-to-end distance probability distribution function of a WLC with $\lambda/L_0 = 1$, as computed numerically from (4.13). The curves denote results at varying levels of truncation of the infinite matrix $\hat{\mathbf{J}}$. *Right panel:* The end-to-end distance probability distribution function $P(r)$, numerically computed from the Fourier transform of $P(q)$. Again, the continuous curves denote results at varying levels of truncation of the infinite matrix $\hat{\mathbf{J}}$. The points are results of a MC calculation performed with 50 segments and 30000 conformations.

4.6.1 *Structure factor of a spherical molecule*

Owing to the rotational symmetry, there is no orientational average; the integration in (4.14) can be readily performed in spherical coordinates with the polar axis chosen along the direction of $\vec{q}$. Performing the integration over the angular variables leaves

$$2 \cdot 2\pi \frac{1}{q^3} \int_0^{qR} dx\, x \sin x = \frac{4\pi}{q^3} (\sin qR - qR \cos qR)$$

where R is the radius of the sphere. Inserting the integral in (4.14) results in

$$S(q) = \left[3 \frac{\sin qR - qR \cos qR}{(qR)^3} \right]^2, \tag{4.15}$$

plotted in Fig. 4.6.

4.6.2 *Structure factor of a cylindrical molecule*

For a cylindrical molecule of length L and radius R the integration in (4.14) proceeds in cylindrical coordinates as follows: the z-axis is chosen parallel to the axis of the cylinder. The vector $\vec{q}$ has components $q_{\parallel} = q\mu$, parallel,

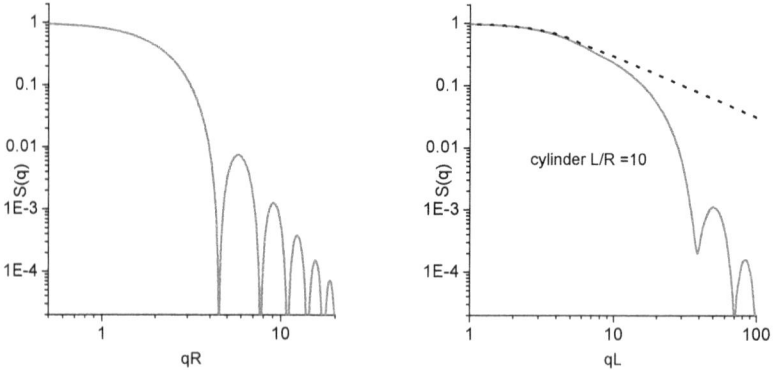

Fig. 4.6   *Left panel:* The structure factor of a sphere of radius R, (4.15) shown as a function of qR in doubly logarithmic scale. *Right panel:* The structure factor of a cylinder of length L and radius $R = L/10$, (4.16), is shown as a function of qL in doubly logarithmic scale. Also shown, as a dotted line, is the limit of the hard rod, (4.17). Note that the cylindrical structure becomes detectable at values $qR \gtrsim 1$. At lower wavenumbers the cylinder is indistinguishable from a thin rod of the same length.

and $q_\perp = q\sqrt{1-\mu^2}$, perpendicular to the z-axis. The integration results in

$$\int_0^R dr\, r \int_{-L/2}^{L/2} dz\, e^{iq_{||}z} \int_0^{2\pi} d\phi\, e^{iq_\perp r \cos\phi} =$$

$$\frac{2\sin(q_{||}L/2)}{q_{||}} \int_0^R dr\, r2\pi J_0(q_\perp r) =$$

$$\frac{2\sin(q_{||}L/2)}{q_{||}} \cdot \frac{2\pi}{q_\perp} \cdot RJ_1(q_\perp R),$$

where $J_{0,1}$ are Bessel functions of the first kind. The structure factor can now be obtained by dividing the last result by the volume of the cylinder, squaring, and performing the orientational average. The resulting integration

$$S(q) = \frac{1}{2}\int_{-1}^{1} d\mu \left[\frac{\sin(q_{||}L/2)}{q_{||}L/2} \cdot \frac{2J_1(q_\perp R)}{q_\perp R}\right]^2 \tag{4.16}$$

must in general be performed numerically. A typical result is shown in Fig. 4.6.

4.6.2.1 *The limiting case of a thin rod*

In the limit $R \to 0$ the cylinder reduces to an infinitely thin rod. Mathematically, this limit is equivalent to a WLC where the ratio λ/L_0 approaches infinity. In this case, $2J_1(x)/x \to 1$, and the integral (4.16) reduces to

$$
\begin{aligned}
S(q) &= \frac{1}{2} \int_{-1}^{1} d\mu \left[\frac{\sin(qL\mu/2)}{qL\mu/2} \right]^2 \\
&= \frac{2}{qL} Si(qL) - \left[\frac{\sin(qL/2)}{qL/2} \right]^2,
\end{aligned}
\tag{4.17}
$$

where $Si(x) = \int_0^x dt \sin t/t$ is the sine integral function. The function (4.17) is shown in Figs. 4.3 and 4.4.

4.7 Light scattering from long DNA molecules

Elastic light scattering experiments have traditionally provided important structural information about macromolecules in solution. This is because macromolecular sizes are often of the order a few hundred nanometers, i.e. comparable with wavelengths of visible light, with a variety of sources to choose from. The earliest studies on DNA [Peterlin (1953a,b)] predate the discovery of the double helix and estimate a persistence length of the order of 30 nm. Subsequent studies [Cohen and Eisenberg (1966); Jolly and Campbell (1972); Godfrey and Eisenberg (1976); Harpst (1980)] have upwards refined this early estimate and studied the detailed dependence of bending stiffness on salt content and sequence length; the temperature dependence of the persistence length has also been studied using light scattering techniques, but this will be discussed separately in 11.

The intensity of the beam scattered in the forward direction, commonly known as the forward scattering intensity, depends (i) on specific, measurable properties of the experimental setup, e.g. incident beam intensity, dielectric constant of the solvent, and (ii) on the total amount of scattering material in the sample. Experimental data on the angular distribution $I(\theta)$ of the scattered intensity are usually reported in the literature in the form

$$
\frac{Kc}{I(\theta)} = \frac{1}{S(\vec{q})} \left[\frac{1}{M} + 2A_2 c + \cdots \right]
\tag{4.18}
$$

for a sequence of samples of decreasing concentration c, from which an extrapolation to the $c \to 0$, $q \to 0$ limits can be performed (*Zimm plots* [Zimm (1948)]). In (4.18), the effects of (i) are included in the measured experimental constant K, M is the molecular weight of the scattering macromolecule,

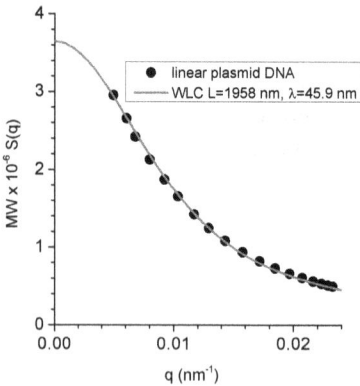

Fig. 4.7 Intensity of elastic light scattering from a linear DNA plasmid composed of 5.76 kbps. The circles denote the experimental data (extracted from the zero concentration limit of [Latulippe and Zydney (2010)]). The continuous curve is a WLC model calculation performed with the nominal length $L_0 = 1958$ nm, a pesistence length $\lambda = 45.9$ nm, and a molecular weight 3.65×10^6.

and the constant A_2, known as the second virial coefficient, provides a measure of intermolecular interactions.

The extrapolated Zimm plots thus provide information about the molecular weight of the solute molecules *and* their structure, as described by the structure factor. In the case of controlled DNA sequences, where the molecular weight is known, this provides a useful further check on the validity of the data and/or the extrapolation procedure.

Fig. 4.7 shows light scattering data [Latulippe and Zydney (2010)] from a 5.76 kbp long linear plasmid DNA in an aqueous solution with 0.2 M salt content. The nominal length of the stretched molecule is thus $L_0 = 5760 \times 0.34 = 1958$ nm and its molecular weight $3.74 \times 10^6 D$ (assuming a value of $650 D/bp$). A WLC model calculation (4.12) with the given L_0 and a persistence length $\lambda = 45.9$ nm provides an excellent fit to the experimental data; moreover, the molecular weight extracted from the absolute intensity, $3.65 \times 10^6 D$, is quite close to the nominal value, adding further credence to the WLC model description.

4.8 Small angle neutron scattering (SANS) from short DNA molecules

Small angle neutron scattering (SANS) can be used to provide structural information on smaller DNA molecules, typically consisting of up to about

200 base pairs. The elastic properties of DNA at this smaller scale are of fundamental interest in view of their relevance to basic biophysical phenomena such as nucleosome wrapping.

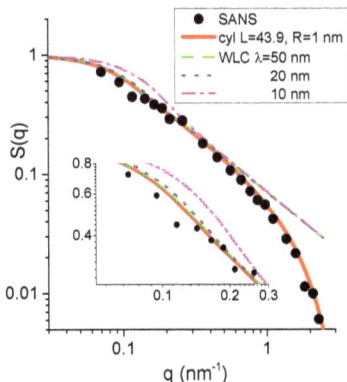

Fig. 4.8   Small angle neutron scattering from a 130 bps long fragment of the T7 phage. The circles represent data from an undeuterated sample [Lederer *et al.* (1986)]. The thick solid curve represents the structure factor of a cylinder, (4.16), of length 44 nm and radius 1 nm. The other three lines represent WLC structure factors computed from (4.12) with $\lambda = 50, 20, 10$ nm, respectively. The inset zooms in the low-wavevector region.

Fig. 4.8 shows some of the possibilities as well as the challenges of neutron scattering methods as applied to small DNA fragments. The scattering data [Lederer *et al.* (1986)] were taken from a 130 bps long fragment of the T7 phage. The data are very well described by scattering from a cylinder of length $L = 44$ nm, equal to the stretched length of the fragment, and $R = 1$ nm, equal to the crystallographically measured radius of the double helix (cf. Eq. 4.16). In other words, at this scale, the overall DNA structure revealed by the neutron scattering experiment is that of a solid cylinder.

Can the experiment provide any further information about the actual stiffness of DNA?

Before discussing that, it is useful to note that the deviation of all three WLC curves from the data at high q does not indicate a deficiency of the WLC model, which is simply not designed to describe the finite thickness of the DNA molecule. Data from the region $q \gtrsim 0.5\,\text{nm}^{-1}$ is beyond the scope of the WLC model. Conversely, any useful statement about flexibility will come from the region where the WLC is, at least a priori, applicable. To this end, Fig. 4.8 includes three WLC curves, computed at $\lambda = 50, 20, 10$ nm, respectively. They are best viewed in the inset. Comparison with the data shows that the available data cannot distinguish between a solid cylinder (more precisely, a hard rod, since, at this wavevector, the thickness of the cylinder is not detectable) and a stiff chain with $\lambda = 50$ or even 20 nm. It *can* however definitively exclude "soft"DNA, i.e. a persistence length of 10 nm.

I will return to the issue of flexibility of short DNA molecules, discussing more recent experimental evidence, in Chapter 12.

Chapter 5

Thermal unbinding of the double helix

5.1 Introduction

5.1.1 *Discovery*

Unbinding of the two DNA strands was observed [Thomas (1954)] in saline solution shortly after the discovery of the double helix. A decrease of the NaCl molar concentration from 10^{-2} to $10^{-4.4}$ was accompanied by an increase in the UV absorption and "denaturation" of the macromolecule, i.e. an irreversible collapse, or "melting" of its secondary structure. In other words, there is no breaking of covalent bonds which form the DNA backbone, but simply an unbinding of the two strands - as evidenced by the molecular weight of the product macromolecule, which is one half of the original. It was soon established that such a strand separation could also be induced by increasing the pH of the solution or by heating the sample beyond a certain temperature. Since temperature is a far more accessible control parameter than salt concentration or pH, the phenomenon of *thermal denaturation of DNA* received significant attention over the following decades.

In this introductory section I will try to give a descriptive account of the phenomenon, using mostly the language of physical chemistry. This will help introduce basic concepts, thermodynamic quantities, and energy scales involved.

5.1.1.1 *Melting profiles; dependence on GC-content*

Since the excess UV absorbance is characteristic of the denatured state, it can be used to measure the fraction θ of unbound base pairs as a function of temperature. Sometimes one may opt to plot the complement, known as

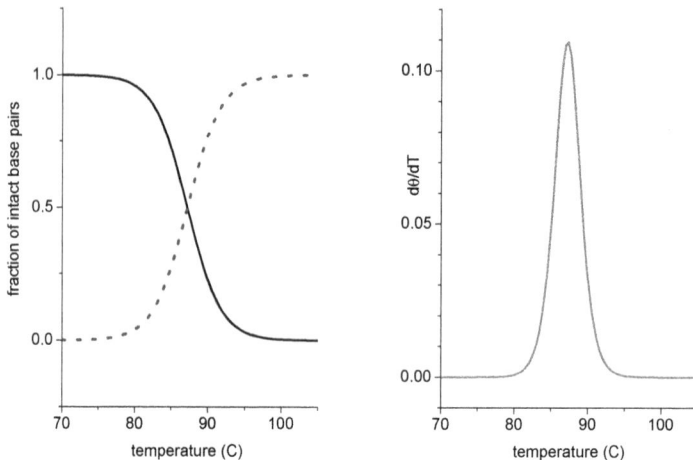

Fig. 5.1   *Left panel:* a schematic DNA melting curve, i.e. the fraction of intact (solid curve) or unbound (dotted curve) base pairs as a function of temperature. *Right panel:* a differential melting curve (profile), i.e. the derivative of the melting curve as a function of temperature. The shapes of both curves are typical of the melting process in a number of chromosomal samples.

the helical fraction. This results in a melting curve, schematically drawn in the left panel of Fig. 5.1. Alternatively, and this will turn out to be instructive in the case where the melting curve exhibits a fine structure, it is possible to plot the temperature derivative of the melting fraction. The differential melting profile for the melting curve of the left panel is shown in the right panel of Fig. 5.1. By convention, the temperature at which one half of the base pairs is still intact is defined as the melting temperature T_m.

Measurements performed on chromosomal samples revealed that the melting temperature increases linearly with GC-content. Some of the original data [Marmur and Doty (1960)] are shown in Fig. 5.2. A cursory look at the structure of the double helix may suggest the higher thermal stability of GC-rich molecules is related to the fact that G-C pairs are bound by 3 hydrogen bonds, whereas A-T pairs only by 2. In fact, as we will see below, this is somewhat misleading, since hydrogen bonds account only partially for the energy that binds the two strands.

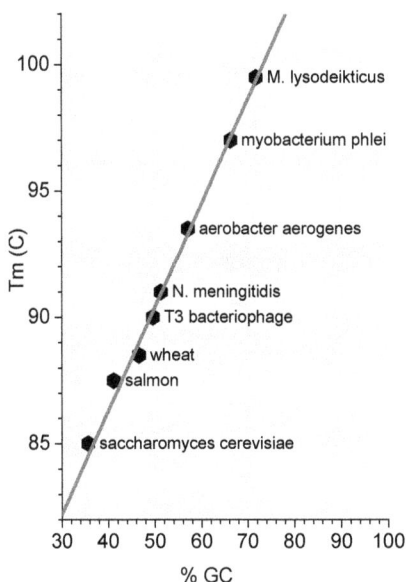

Fig. 5.2   Melting temperature of DNA vs. GC content of selected chromosomal samples [Marmur and Doty (1960)] in an aqueous solution with 0.2M Na$^+$ concentration.

5.1.2   Basic Thermodynamics

5.1.2.1   Oligonucleotide duplex melting. Helix-Coil Equilibrium

Synthetic oligonucleotide duplexes, i.e. molecules consisting of a few base pairs, have proved essential for understanding DNA thermodynamics. One of the reasons is that many of them melt in an "all or none" fashion. Either the double helix is intact with all base pairs bound in a molecule, or the strands have completely dissociated. However, in a solution, all molecules do not dissociate simultaneously. At any given temperature there is a coexistence of ordered, double-helical and disordered, random-coil-like strands. The coexistence of double-stranded and single-stranded DNA can be understood as a chemical equilibrium between two species

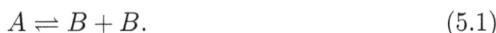

$$A \rightleftharpoons B + B. \tag{5.1}$$

with an equilibrium constant

$$K \equiv \frac{[B][B]}{[A]} = \frac{[c(1-\theta)]^2}{c/2\theta} \equiv e^{-\Delta G/k_B T} = e^{\Delta S/k_B} e^{-\Delta H/k_B T} \tag{5.2}$$

where c is the total strand concentration, θ the fraction in the double strands, and $\Delta G = \Delta H - T\Delta S$ is the transition free energy, i.e. the difference in free energy between the two species, consisting of an enthalpic and an entropic part. The thermodynamic stability of the double helix is governed by ΔG.

It should be noted that - as befits a chemical equilibrium between two species - the transition from double- to single-stranded oligonucleotides is *reversible*, in contrast to what happens in natural DNA samples. This difference can readily be understood in terms of the far greater complexity of the kinetic pathways involved in successful strand matching of natural samples. As we shall see below, the detailed study of the denaturation transition in oligonucleotides provides a unique insight into the thermodynamic stability of DNA.

5.1.2.2 Van't Hoff vs. calorimetric enthalpies. Cooperativity of the transition.

At $T = T_m$, $\theta = 1/2$ and (5.2) can be written as

$$\frac{1}{T_m} = -\frac{k_B}{\Delta H} \ln c + \frac{\Delta S}{\Delta H}. \qquad (5.3)$$

Consequently, plotting the inverse of the melting temperature vs. the logarithm of the concentration can be used to estimate ΔH and ΔS.

If the temperature dependence of the enthalpy ΔH and the entropy ΔS of the transition can be neglected (a good assumption for many oligonucleotides) then it is straightforward to show that

$$\left(\frac{d\theta}{dT}\right)_{\theta=1/2} = -\frac{\Delta H}{6k_B T_m^2}. \qquad (5.4)$$

In other words, the slope of the melting curve at $T = T_m$ (taken, in principle, at any concentration) may also be used to determine the enthalpy of the transition.

How is this enthalpy, computed from the UV absorbance melting profile, and known as the *van't Hoff enthalpy*, related to the one determined directly, e.g. by calorimetric means? In general, for a transition which involves intermediate states, (5.2) does not hold and therefore there is no reason for the two quantities to be equal. However, if the transition is of the "all or none" type, i.e. no intermediates are involved, *and* ΔH and ΔS do not depend strongly on temperature, then the argument sketched above is valid, so that van't Hoff and calorimetric enthalpies (and entropies) will be equal.

Mutatis mutandis, the equality of the two enthalpies can be used as a test of the "all or none" property. Or, at a more qualitative level, comparison of the two enthalpies can be used to estimate the degree of "cooperativity" of the transition. The closer the van't Hoff enthalpy is to the calorimetric one, the more "cooperative" the helix-coil transition, i.e. the higher the fraction of the base pairs which melt simultaneously.

5.2 Base sequence and thermodynamic stability

5.2.1 *The nearest-neighbor model*

Systematic analysis of oligonucleotide duplex melting by a number of research groups during the 1980's and the 1990's revealed a relationship between base sequence and thermodynamic stability which is more complex than the simple linear relationship between melting temperature and GC-content might suggest. In other words, it has been shown that is not possible to "assign" individual thermodynamic enthalpies and/or entropies to the two types of Watson–Crick pairs in a way that allows the predictive computation of the total ΔG for any given oligonucleotide.

What has been possible, after an exhaustive analysis of hundreds of different oligonucleotides [Breslauer *et al.* (1986); SantaLucia (1998)], is to validate the so-called nearest-neighbor model, according to which the thermodynamic stability of a given base pair depends on the identity and orientation of neighboring base pairs. In other words, individual transition enthalpies and entropies can be assigned to distinct types of neighboring base pairs. Using these as a standard set, it is possible to predict the total ΔG for any given oligonucleotide duplex. Moreover, making use of the computed ΔG it is possible to predict the melting temperature of the oligonucleotide duplex, using (5.3).

5.2.1.1 *Dimer notation*

In the duplex oligonucleotide

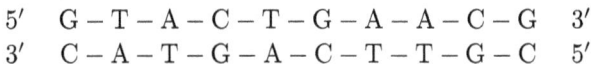

$$5' \quad G - T - A - C - T - G - A - A - C - G \quad 3'$$
$$3' \quad C - A - T - G - A - C - T - T - G - C \quad 5'$$

the dimer consisting of e.g. the fourth and fifth Watson–Crick base pairs C∘G, T∘A will be collectively denoted as CT/GA. Reading the lower line in the 5'3' and the upper line in the 3'5' direction, the same dimer would be denoted as AG/TC. Interaction properties however are the same for these two pairings. Conversely, the dimer consisting of the second and third Watson–Crick pairs reads TA/AT in both cases. Because of six such reductions of equivalent nearest neighbor sequences, there will only be a total of $4^2 - 6 = 10$ distinct pairwise interactions.

The nearest neighbor model sets out to express the transition free energy for *any* oligonucleotide duplex as a sum over nearest-neighbor pair free energies plus the corresponding initiation and termination terms

$$\Delta G_N = \Delta G_1^{init} + \sum_{i=1}^{N-1} \Delta G_{i,i+1} + \Delta G_N^{init}, \tag{5.5}$$

where the $\Delta G_{i,i+1}$ can take any of ten values, characteristic of the two neighboring Watson–Crick pairs. In addition, there are correction terms to account for the end base pairs.

5.2.2 *Enthalpies and entropies for neighboring base pairs*

The transition enthalpies and entropies corresponding to distinct neighboring pairs (dimers), as determined in [Allawi and SantaLucia (1997)] for a salt concentration of 1M, are shown in Table 5.1. In addition, in order to give a measure of thermodynamic stability under physiologically relevant conditions, the value of $\Delta G = \Delta H - T\Delta S$ is given at a temperature of 37 C for each neighboring pair.

Table 5.1   Nearest neighbor enthalpies and entropies at 1 M NaCl (from [Allawi and SantaLucia (1997)]). Error estimates are given in parentheses. The last two rows list correction terms for initial and final base pairs.

Sequence	ΔH (kcal/mol)	ΔS (cal/K/mol)	ΔG_{37} (kcal/mol)
AA/TT	7.9 (0.2)	22.2 (0.8)	1.00 (0.01)
AT/TA	7.2 (0.7)	20.4 (2.4)	0.88 (0.04)
TA/AT	7.2 (0.9)	21.3 (2.4)	0.58 (0.06)
CA/GT	8.5 (0.6)	22.7 (2.0)	1.45 (0.06)
GT/CA	8.4 (0.5)	22.4 (2.0)	1.44 (0.04)
CT/GA	7.8 (0.6)	21.0 (2.0)	1.28 (0.03)
GA/CT	8.2 (0.6)	22.2 (1.7)	1.30 (0.03)
CG/GC	10.6 (0.6)	27.2 (2.6)	2.17 (0.05)
GC/CG	9.8 (0.4)	24.4 (2.0)	2.24 (0.03)
GG/CC	8.0 (0.9)	19.9 (1.8)	1.84 (0.04)
G∘C init	-0.1 (1.1)	2.8 (0.2)	- 0.98 (0.05)
A∘T init	-2.3 (1.3)	-4.1 (0.2)	-1.03 (0.05)

The following comments are appropriate at this point:

5.2.2.1 *Sign convention*

Both ΔH and ΔS are positive. They have been defined as transition enthalpy and entropy, respectively, for duplex melting. The first expresses the energy difference between the product (two single strands) and the input (the duplex). The difference is in essence the binding energy of the helix. It is what a calorimetric experiment would measure, integrating the specific heat over a temperature range that includes T_m. The second is the entropy difference between the two single strands and the duplex. The transition is entropically favorable because the duplex is a much more ordered state than the single strands.

5.2.2.2 *Magnitude of ΔH*

Dimer transition enthalpies vary between 7.2 and 10.6 kcal/mol. They are clearly higher when the dimer consists solely of G∘C and/or C∘G base pairs and lower if the dimer consists solely of A∘T and/or T∘A base pairs. How does this compare with typical hydrogen bonding energies in aqueous solutions? For comparison, the latent heat of water vaporization is about 9.7 kcal/mol. It is estimated that vaporization from the bulk involves the breaking of 4 hydrogen bonds, which would result in a rough estimate of 2.4 kcal/mol per hydrogen bond. More detailed estimates based on computational studies and spectroscopic as well as thermodynamic data suggest a value of 1.9 kcal/mol [Silverstein *et al.* (2000)]. Allowing for 2 or 3 hydrogen bonds for each A∘T or G∘C pair respectively would only account for roughly half the dimer transition enthalpy. This is consistent with the idea of strong *stacking* interactions originating in the $\pi-$ electrons of successive, parallel, planar base pairs.

5.2.2.3 *Magnitude of ΔS*

The dimer transition entropies listed in Table 5.1 vary from 19.9 to 27.2 cal/K/mol. The average value is 22.4 cal/K/mol with a standard deviation of 2.1 cal/K/mol. In units more appropriate to statistical mechanics, this means that melting of the duplex produces on the average an entropy of 11.3 k_B per base pair. Similarly, Delcourt and Blake [Delcourt and Blake (1991)], concluded that the cooperative melting behavior of a large number of samples with varying sequence and composition can be consistently described by the constant value $\Delta S \approx 12.5(\pm 0.9)k_B$.

The above values of the DNA melting entropy — which are mutually consistent within their respective error bars — can be compared with the average gain in entropy from breaking a hydrogen bond, which has been estimated [Silverstein *et al.* (2000)] to be about 2.4 k_B. This would imply a contribution to the melting entropy of G∘C and A∘T pairs, respectively, of 7.2 and 4.8 k_B; on average therefore, about half the melting entropy appears to be accounted by the breaking of hydrogen bonds.

5.2.2.4 *Thermodynamic stability*

The thermodynamic stability of each dimer at any given temperature is governed by ΔG. The values at 37 C vary from 0.58 for the TA/AT to 2.24 kcal/mol for the GC/CG dimer. If we compare this with the thermal energy $k_B \times 310$ K = 0.616 kcal/mol, it becomes clear that the TA/AT dimer is very marginally stable. The same holds true, to a somewhat lesser extent, for AT/TA and AA/TT dimers. Regions of the double helix which contain these motifs will tend to be dominated by thermal fluctuations, resulting an a locally open structure. In fact even the relatively stable GC/CG dimer has a ΔG which is only 3.6 times the thermal energy, so the probability of its opening spontaneously will not be negligible.

5.2.2.5 *Helix initiation and termination*

The nearest neighbor parameters [Allawi and SantaLucia (1997)] also contain appropriate corrections for the end terms. The terms are assumed to be symmetric, i.e. initiation (or termination) of the helix with a G∘C pair is equivalent to initiation (or termination) with a C∘G pair. Note that their relative stability is negative, i.e. there is a free energy cost for maintaining the ends of the double helix at 37 C.

5.2.2.6 *Dependence on salt concentration*

The parameters of Table 5.1 have been derived for a particular NaCl concentration. Melting temperatures of oligonucleotides can be predicted with an accuracy of about 2 C if one assumes that the transition enthalpies are independent of salt concentration while transition entropies vary as

$$\frac{\partial \Delta S}{\partial \log_{10}[c]} = 0.848N \text{ cal/mol} \tag{5.6}$$

where N is the number of base pairs and c the molar concentration of Na^+ ions. [SantaLucia (1998)].

5.2.2.7 *Melting of long sequences*

The empirical Marmur–Schildkraut–Doty [Marmur and Doty (1960); Schildkraut *et al.* (1962); Blake and Delcourt (1998)] equation relates the melting temperature

$$T_m(C) = 193.67 - (3.09 - x) \cdot (34.47 - 6.52 \log_{10}[c]) \qquad (5.7)$$

of a long sequence ($N > 1000$) to the GC fraction x and the molar concentration, c, of Na^+ ions.

5.3   Melting of longer DNA chains

5.3.1   *Internal vs. external melting*

As sequences become longer, the all-or-none model of melting ceases to be applicable. The denaturation process becomes more complex. As the temperature is raised, each duplex follows its own path to the unfolded, coil-like state; groups of spatially close base pairs become unbound so that many duplexes are only partially melted.

In order to describe what happens, we have to distinguish between two fractions, or probabilities. First, the probability that a duplex has at least one bound pair, conventionally denoted by θ_{ext}, and expressing the fraction of double strands which remain connected. Second, the conditional probability, for a duplex which is a member of that class, i.e. having at least one bound base pair, that any given base pair belonging to it will also be bound. The second quantity is conventionally denoted by θ_{int} and measures the fraction of intact base pairs in a duplex that has at least one intact base pair. The underlying assumption — which makes the distinction between external and internal melting fractions useful — is that the latter depends solely on the average properties of single chains.

The total fraction of bound base pairs, which is the quantity detected by UV absorbance or calorimetry, will be given by the product

$$\theta = \theta_{ext}\theta_{int}.$$

The all-or-none approximation is equivalent to setting $\theta_{int} = 1$ at all temperatures. The reason why it works so well in oligonucleotides is that DNA melting is a highly cooperative process. For a short sequence cooperativity is in this sense complete. Unbinding of all pairs occurs simultaneously.

The measurement process detects $\theta \approx \theta_{ext}$. At the opposite end, very long chains, containing a few thousands of base pairs, will retain some islands of stability until the very end of the melting process. In this case, the value of θ_{ext} will remain very close to 1, until the melting process is almost complete. Therefore, for all practical purposes, for melting in long chains (meaning in practice $N > 500$), $\theta \approx \theta_{int}$ will be an excellent approximation.

In view of the above, sequences of intermediate length (practically meaning $20 < N < 500$) present some serious difficulties. Theoretical work, applying methods of statistical mechanics, and which will be described in the next chapters, can achieve a reasonably good understanding of θ_{int} for any length; there is, however, no generally applicable theory to describe θ_{ext}. Furthermore, there was, until very recently, no generally applicable experimental procedure that measures θ and θ_{ext} simultaneously. In view of the above, a full interpretation of experimental data for chains of intermediate length remains elusive.

In the remainder of this chapter I will give a brief description of a few selected examples of melting in longer chains, along with the theoretical challenges they present.

5.3.2   *Multistep melting*

As described in section 5.1.1 early results on DNA melting obtained from natural samples revealed no internal structure in the melting profiles. In contrast, work with viral and bacterial DNA very soon revealed that melting proceeded in steps. A schematic illustration of 3-step melting is provided in Fig. 5.3.

Multistep melting is consistent with the linear dependence of T_m on GC-content. As the temperature is raised, AT-rich regions tend to melt first, leaving GC-rich ones intact until a somewhat higher temperature is reached. In fact, the property of multistep melting is an intrinsic feature of heterogeneity; the only reason it is not detected in samples of natural DNA is that the statistics over millions of base pairs average out individual steps and produce what appears to be a single, rounded peak.

5.3.3   *Melting of long, synthetic, homogeneous duplexes*

Heterogeneity — along with the associated multistep melting properties — can be eliminated by studying polynucleotides, i.e. long polymer chains consisting of identical base pairs. It has been possible to synthesize this kind of "homogeneous DNA" up to lengths of order of 10-20 thousand of base

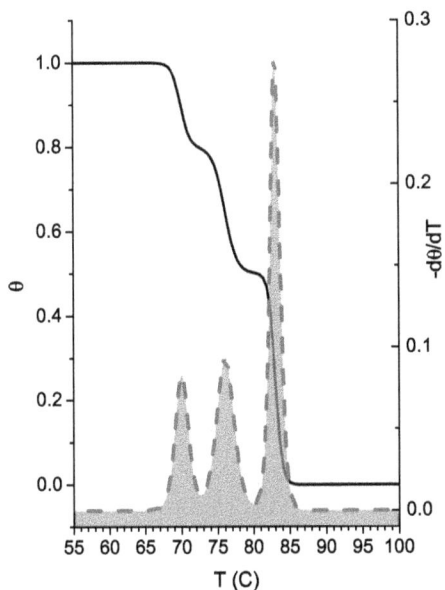

Fig. 5.3 Schematic multistep melting. The helical fraction is denoted by the solid line, the negative of its temperature derivative (right y-axis) by the dashed line.

pairs. The resulting melting profiles [Inman and Baldwin (1964)] can be extremely sharp, with typical transition widths $\Delta T = -T_m \left(d\theta/dT \right)_{\theta=1/2}$ of the order of one degree or less. Such a finite width could in principle be attributed to the finite size of the chain. Unfortunately, to my knowledge, there are no systematic studies of the dependence of the transition width on chain size to support such a conjecture on a *purely experimental* basis. On the other hand, *mathematical models* of the DNA melting, to be presented in the following sections, do exhibit a sharp thermodynamic phase transition in the limit $N \to \infty$.

Melting of long homogeneous duplexes has been observed to be largely reversible [Inman and Baldwin (1964)]. Again, as with oligonucleotides (cf.above), this is consistent with the decreased complexity of the kinetic pathway involved in strand recombination, compared to genomic samples.

Chapter 6

Mechanical unbinding of the double helix

6.1 Unzipping

6.1.1 The experimental findings

An altogether different way of *mechanically* unbinding double-helical DNA has been accomplished via literally pulling (*"unzipping"*) its two strands apart. A somewhat simplified, schematic arrangement of the original experiment[Essevaz-Roulet *et al.* (1997)] is shown in Fig. 6.1. One strand is attached to a sliding plane, while the other is attached to metal ball via covalent bonding. In the original experiment the sliding plane moves at a constant velocity, which translates to a linearly growing displacement of the strand end, while the force is measured on the other end via a needle attached to the ball. The displacement control parameter provides therefore, indirectly, a measure of the end-to-end distance of the unbound portion of the DNA, while the force acting on the one end is the response to the sliding motion. Measurements of force vs. displacement show roughly a plateau at a force around 13 pN extending over a displacement interval which corresponds to the full transition to an almost stretched, unzipped, single-stranded DNA. The measured force exhibits a noisy behavior around the mean plateau value, with an amplitude of about $2-3\,pN$. The fluctuations are however *not* stochastic; a careful analysis shows that they follow the variation of GC vs. AT composition of the sample.

A detailed understanding of how local openings of the double helix occur under the action of a force in the pN range and their dependence on sequence variation is highly relevant to the study of DNA replication.

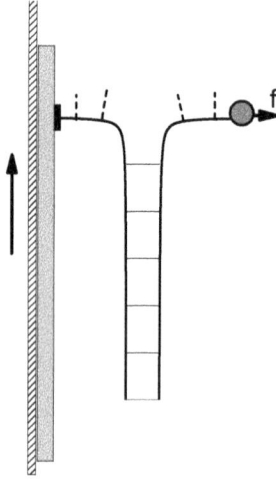

Fig. 6.1   A schematic representation of the DNA unzipping experiment[Essevaz-Roulet *et al.* (1997)]. One end of the double-helix is attached to the sliding plane, the other to a metal ball. The force acting on the second end is measured via a needle attached to the ball. The displacement of the sliding plane provides a measure of the end-to-end distance of the unbound part of the chain (single-stranded DNA). The figure shows a chain with a total $N = 7$ base pairs, $m = 2$ of which are unbound.

6.1.2   *Mean-field theory*

Following [Lubensky and Nelson (2000, 2002)], it is possible to formulate a mean-field theory of unzipping in a statistical ensemble corresponding to constant force. This does not correspond exactly to the experimental situation, but it will serve for a basic understanding of the phenomenon.

Consider the partially open chain depicted in Fig. 6.1. It can be thought of as being composed of two parts. First, the portion of intact, double-stranded DNA, composed of $(N - m)$ base pairs. Second, the open, single-stranded portion, composed of $2m$ bases; the force f acts on the end site of this open part.

The equilibrium of the two parts will be determined by the minimum of the total free energy. The double-stranded part has a free energy that can be estimated using the material presented in the previous chapter. In particular, it was shown in section 5.2.2 that thermodynamic free energies depend on the sequence (dimer contributions) rather than simply the GC content. Nonetheless, for our purposes, it is sufficient to use a mean-field estimate for the free energy per bound pair

$$g_0 = \Delta S(T - T_m) \tag{6.1}$$

where $\Delta S \approx 12.5 k_B$ is the average melting entropy of a base pair (cf. Section 5.2.2) and T_m the melting temperature of the sample. The latter can be estimated for the λ-phage used in the unzipping experiment to be approximately 90°C. This gives a room temperature estimate of $g_0 = -3k_BT$.

I now turn my attention to the open, single-stranded part. Given the short persistence length of ss-DNA (of the order of the monomer distance), the use of a continuum model is ill-advised. A quick way to proceed is by exploiting the analytically tractable FJC model (cf. 2.2) with an effective monomer (Kuhn) length of $\ell = 2.6\ nm$, equal to twice the commonly estimated persistence length of 1.3 nm. The free energy per base at an external force field f^1 is then given by

$$g_1 = -k_BT\frac{a'}{\ell}\ln\left[\frac{\sinh(\beta\ell f)}{\beta\ell f}\right], \tag{6.2}$$

where $a' = 0.63nm$ is the crystallographically estimated monomer distance for ss-DNA.

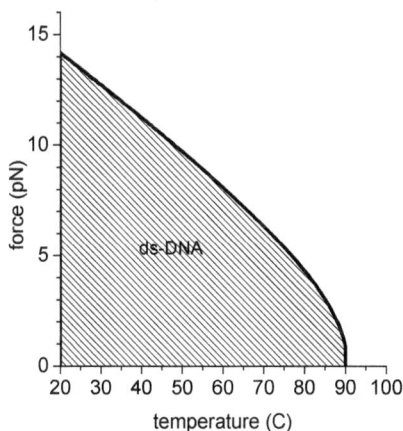

Fig. 6.2 The thermodynamic stability diagram for unzipping, according to (6.3). The force necessary to bring about unzipping is plotted as a function of temperature. Double-stranded DNA is stable inside the shaded region.

---

[1]There is no stretching elastic energy associated with the double-stranded part of the molecule since, as one can see from the schematic of Fig. 6.1 the force is transverse to the double-stranded part.

The equilibrium of the system, composed of $N - m$ bound base pairs and $2m$ unbound bases, will be governed by the minimum of the total free energy

$$G(m) = (N - m)\, g_0 + 2mg_1. \tag{6.3}$$

In fact, since $G(m)$ is just a linear function of m, thermodynamic stability will depend on the sign of $-g_0 + 2g_1$. As long as this quantity is positive, any $m > 0$ will be thermodynamically unstable. Conversely, if $-g(0) + 2g(1) < 0$, spontaneous unzipping will occur with the maximum possible number $m = N$. The resulting thermodynamic phase diagram is shown in Fig. 6.2. At room temperatures, unzipping is predicted to occur at about 14 pN, very close to the observed value (13 pN).

6.2   Overstretching

6.2.1   *The overstretching transition*

It has been possible [Cluzel *et al.* (1996); Smith *et al.* (1996)] to extend direct manipulation of the DNA molecule to longitudinal forces beyond the few pN described in Section 2.5. As shown in Fig. 6.3, when the applied force reaches a level of about 65 pN, the force-extension curve exhibits a plateau. The plateau extends until the contour length of the DNA chain "overstretches", i.e. increases to about 1.7 times the original length; beyond that contour length, an elastic response sets on again.

The extension by a factor of about 1.7 corresponds roughly to the ratio of monomer distances of single-stranded and double-stranded DNA. It would be therefore natural to interpret the transition as a conversion from ds- to ss-DNA. Owing however to the geometry of the experimental setup, the two strands do not come entirely apart; their proximity to each other during and after the transition renders the latter to a large extent reversible. Therefore, the transition was originally attributed [Cluzel *et al.* (1996)] as a conversion from the double helical B-DNA to a new ladder-like form, S-DNA, in which base pairs unstack, but hydrogen bonding - and thus base pairing - persists.

In Section 6.2.3 I will show, following [Rouzina and Bloomfield (2001)], that a careful thermodynamic analysis of the stretching curve is entirely consistent with the intuitive idea of force-induced melting, in other words, with the conversion of ds- to ss-DNA. This interpretation has been recently confirmed by further experimental work [van Mameren *et al.* (2009)] in which the percentage of ds-DNA was directly detected and measured to

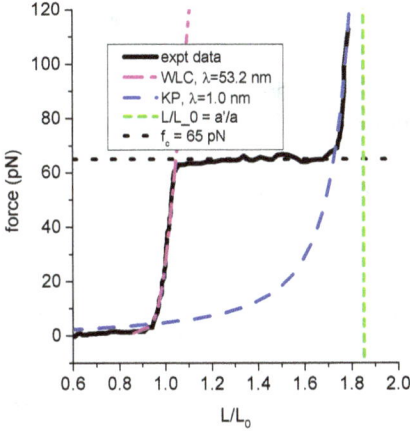

Fig. 6.3   The DNA force-extension curve at forces up to 120 pN. Extension is measured relative to the ds-DNA contour length. Continuous line, experimental data redrawn from [van Mameren *et al.* (2009)]. Dashed-dotted curve, theoretical WLC calculation according to (2.13), with $\lambda = 53.2$ nm, contour length $L_0 = 8400 \times 0.34$ nm, and corrected as in (6.4) with an elastic (stretch) modulus $F_0 = 1000$ pN. Dashed curve, discrete Kratky–Porod calculation according to (2.24), with $N = 8400$, monomer distance $a' = 0.63$ nm, stiffness constant $\kappa/a' = 8$ pN nm, corresponding to $\lambda = 1.0$ nm, and corrected for elastic stretching, according to (6.4), with modulus $F_0 = 8.4nN$[Hugel *et al.* (2005)]. A dotted horizontal line at 65 pN denotes the overstretch threshold. A vertical, short-dashed line shows the limit of ss-DNA contour length. (from [Theodorakopoulos (2019)])

decrease linearly along the plateau from 100 to 0% , in full accordance with the picture of force-induced melting.

6.2.2   *Enthalpic corrections to the force-extension curve*

As the applied force increases beyond a few pN, hitherto neglected standard mechanical restoring forces come into play. In the case of double-stranded DNA, π-electron stacking forces provide a common physical origin of the stretching and bending modulus; enthalpic and entropic elasticity are both important near the fully stretched state. The estimated [Bustamante *et al.* (2000); Marko and Cocco (2003)] stretch modulus is about $F_0^{ds} = 1000$ pN.

For covalently bonded polymers this threshold may not set in until very high forces are applied. Early AFM measurements performed in the case of ss-DNA, which is bound by covalent bonds, show that it can sustain forces at least up to 800 pN, with an elastic stretch of the order of 15% beyond the crystallographic monomer distance [Clausen-Schaumann *et al.* (2000)].

This suggests an elastic stretch modulus of the order of 5 nN, implying a correction of the order of 1% in the context of a 100 pN experiment. Further work combining AFM measurements with *ab initio* calculations [Hugel *et al.* (2005)] estimates a value of the elastic modulus $F_0^{ss} = 8.4$ nN for ss-DNA. I will use this estimate [Theodorakopoulos (2019)] in the following section.

In the context of the force-extension relationship, elastic stretching can be explicitly taken into account by the correction factor

$$\left(\frac{L}{L_0}\right)_{total} = \frac{L}{L_0} \cdot \left(1 + \frac{f}{F_0}\right), \tag{6.4}$$

where the first factor originates from the entropic elasticity. The corresponding elastic energy per monomer unit

$$\frac{1}{2}\frac{a}{F_0^{ds}}f^2, \tag{6.5}$$

where a is the average distance between monomers, should be included in free energy calculations.

6.2.3   *Force induced melting: the torsionally unconstrained case*

Fig. 6.4 illustrates the standard setup of an overstretching experiment in its "canonical" version, where the duplex is attached at both its 3' 3' ends and therefore both putative single strands resulting from the process would be torsionally unconstrained. Thermodynamic analysis of the plateau in the force-extension curve is in this case straightforward. Melting of a base pair decreases the elastic energy of the double-stranded part by an amount g_{ds} and increases the elastic energy of the attached portion of single strands which is under tension by an amount g_{ss}. The contributions per monomer (base-pair and base, respectively, for the ds- and the ss- chain) must be computed separately.

For long double-stranded DNA chains, the WLC has been shown to be an excellent approximation (cf. Fig. 2.2). We therefore make use of (2.11), corrected for mechanical (enthalpic) stretch elasticity, which gives

$$g_{ds} = k_B T \frac{a}{\lambda} \Lambda_0 \left(\frac{\lambda f}{k_B T}\right) + \frac{1}{2}\frac{a}{F_0^{ds}}f^2 \tag{6.6}$$

for the elastic part of the free energy per base pair in terms of the applied force f and the persistence length λ. The largest eigenvalue Λ_0 of the matrix **J** can be computed numerically (cf. 2.3.2.3). Regarding single-stranded DNA there are two options. If we are interested in an order-of-magnitude

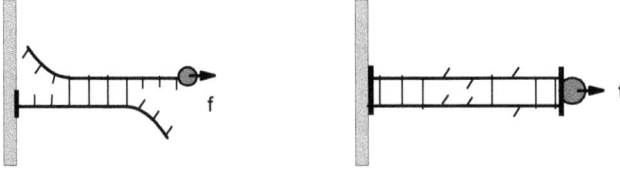

Fig. 6.4 Schematic views of the overstretching set up [van Mameren *et al.* (2009)]. *Left panel:* the "canonical" case. The duplex is attached at the 3' 3' ends. Melting is initiated at the ends and both single strands are torsionally unconstrained. *Right panel:* The 3' 5' 5' 3' attachment geometry. All ends are attached. Melting can be initiated at any point, therefore the DNA consists of both ss- and ds-parts which are all torsionally constrained.

estimate, the analytical expression (6.2) provided by the FJC model is quite acceptable. On the other hand, in view of the high flexibility of ss-DNA, a correct description of the elastic free energy demands the use of the discrete KP model with a free energy per monomer. The second option uses the numerically computed KP elastic free energy per base

$$g_{ss}^{(2)} = -k_B T \ln \mu_0 + \frac{1}{2} \frac{a}{F_0^{ss}} f^2, \tag{6.7}$$

with parameters appropriate for ss-DNA, $J' = 8\,pN \cdot nm$, $a' = 0.63\,nm$, corresponding to a persistence length of $1.0\,nm$. The first option uses the FJC expression (cf. previous section)

$$g_{ss}^{(1)} = -k_B T \frac{a'}{\ell} \ln \left[\frac{\sinh(\beta \ell f)}{\beta \ell f} \right] + \frac{1}{2} \frac{a}{F_0^{ss}} f^2, \tag{6.8}$$

where the segment length $\ell = 2\lambda$ is taken as twice the persistence length. In both options the second term accounts for the mechanical stretch energy.

The resulting difference in the elastic free energy $g_{ss} - g_{ds}$ is shown in Fig. 6.5. The coexistence of the two types of DNA, single-stranded and double-stranded will occur when the difference in the elastic free energy between them achieves full compensation of the duplex stability free energy $g_0 = (T - T_m)\Delta s^*$ (cf. previous discussion of the unzipping case). The condition

$$g_{ss} - g_{ds} = g_0 \tag{6.9}$$

will be satisfied when the force reaches $64\,pN$, according to the second option, in full agreement with experiment. The FJC approximation predicts a slightly lower overstretching force.

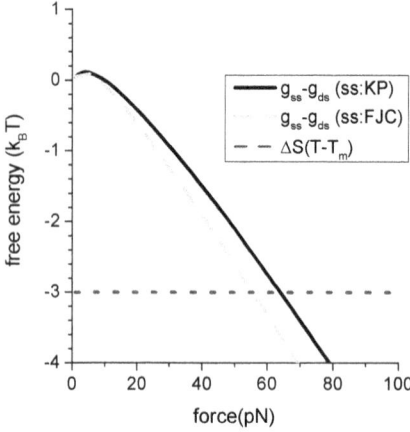

Fig. 6.5   Thermodynamic analysis of the DNA overstretching plateau in terms of force-induced melting [Rouzina and Bloomfield (2001); Theodorakopoulos (2019)]. The difference in the elastic energies $g_{ss} - g_{ds}$, in units of $k_B T$, is plotted as a function of the applied force for the case of the unconstrained geometry described in Fig. 6.4. Both variants of the calculation of g_{ss} are shown, the full numerical discrete KP (full curve) and the approximate FJC-based (dashed curve). Also shown (dotted line) is the constant value $g_0/(k_B T)$. The intersection occurs at $f = 64 \, pN$ in the case of the KP-based curve, and at a slightly smaller value in the FJC case.

6.2.4   *Torsionally constrained forced-induced melting*

DNA overstretching can also be performed in a different attachment geometry (cf. right panel of Fig. 6.4); when all ends are attached, the polymer chain is torsionally constrained. The overstretching transition, again to a length roughly 1.7 times the natural contour length of ds-DNA, now occurs at a significantly higher force, $f = 110 \, pN$ [van Mameren *et al.* (2009)].

The analysis in terms of force-induced melting proceeds along similar lines as in the previous case, with two important differences:

- first, along the force-extension plateau, there are now effectively *two* stretched ss-stranded chains with m bases each (as well as a ds-chain with $N - m$ base pairs). Each of the ss-chains is subjected to one half of the stretching force. Therefore, the elastic free energy advantage of force-induced melting per base pair will be

$$2g_{ss}(f/2) - g_{ds}(f).$$

Compensation of the full duplex stability free energy will now occur at a somewhat higher force, around $80\,pN$ [Rouzina and Bloomfield (2001)] (cf. Fig. 6.7).

- second, the geometry of the constraint prohibits a full entropy release in the overstretched state. The two strands remain stretched, attached at both ends. Moreover, available conformational space is further restricted by the entanglement brought about by the previous double helical order. An "inherited" structure seems to persist in the overstretched state for which "the most straightforward explanation ... is that the new structure, generated during overstretching at 110 pN, consists of two single DNA strands lacking hydrogen bonds between the bases, wrapped around each other with a linking number close to that of relaxed dsDNA"[van Mameren *et al.* (2009)].

The quantity of interest in this context is the entropy difference between the molten state with the geometrical constraint and the one without it, or, in other words, the *entropic penalty* $\Delta S^* < 0$ required to maintain the molten DNA with the inherited structure, compared to the reference state of the two free strands. Geometrically constrained force-induced melting does not release the full amount of DNA melting entropy $\Delta S = 12.5\,k_B$, but a reduced amount $\Delta S + \Delta S^*$. This modifies the relative thermodynamic stability of the double helix (6.1), which now reads

$$\Delta G = (T - T_m)\Delta S + T\Delta S^* \tag{6.10}$$

per base pair, and results in an *enhanced stability* of the double helix. Overstretching will now occur when the elastic free energy difference compensates (6.10).

6.2.4.1 *Estimation of the entropic penalty [Theodorakopoulos (2019)]*

The overstretched DNA with the inherited, plectonemic structure can be visualized as consisting of the two strands joined at successive points with a spacing $10a$, equal to the pitch of the original double helix.

Neglecting any effects arising from excluded volume, the probability of two distinct, partially stretched strand pieces, each with ν elements, starting from a common origin and meeting at a certain point $\vec{r}$ in space within an axial distance b and a radial distance ρ, is p_ν^2, where

$$p_\nu = \pi b \rho^2 \, P_\nu(\vec{r}) \tag{6.11}$$

and $P_\nu(\vec{r})$ is the vector end-to-end distance probability (cf. 1.4.0.1). The conformation with a total of $N/10$ plectonemic joints will have a probability of occurrence $(p_{10}^2)^{N/10}$, i.e. the corresponding entropic penalty per base pair will be equal to

$$\Delta S^* = k_B \frac{1}{5} \ln p_{10}.$$

For a KP chain with $\nu = 10$ and the ss-DNA parameters used in the force-extension curve, Fig. 6.3, this quantity can be computed directly from (1.36) by matrix multiplication and Fourier inversion [Yan *et al.* (2005); Errami *et al.* (2007)]. The numerical results for $P_{10}(\vec{r})$ are shown in Fig.6.6. At the point of interest $r = 10a$, $r/a' = 10a/a' = 10 \times 0.34/0.63 = 5.4$, $P_{10} = 5.5 \times 10^{-4}/(a')^3$. Using a rough estimate of what constitutes a contact between the two strands, $b = \rho = a/2$, results in an entropic penalty

$$\Delta S^* = -2.06 \, k_B$$

per original base pair.

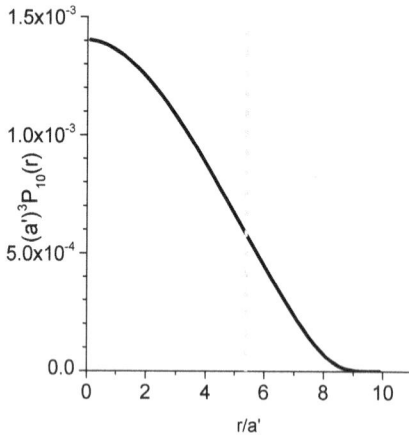

Fig. 6.6   The room temperature vector end-to-end distribution (1.36) for a KP model chain with 10 segments, stiffness constant 8 pN nm; the values correspond to ss-DNA. If the monomer distance is 0.63 nm, the persistence length is 1.0 nm. A vertical line is drawn at a distance $10a$, equal to the original pitch of the double helix.

6.2.4.2   *Determination of the force threshold*

The free energy balance in the case of constrained geometry is shown in Fig. 6.7. It occurs when the difference in free energies between single- and double-stranded state compensates the stability free energy of the double helix, as corrected for the entropic penalty arising from the constraint, (6.10). The theoretically calculated overstretching force threshold, 111 pN, is in excellent agreement with the experimentally observed value, 110 pN.

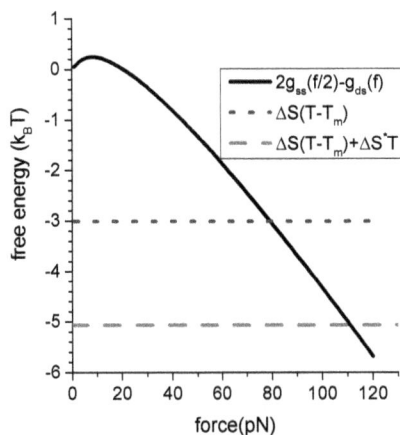

Fig. 6.7 Free energy balance resulting in DNA overstretching (constrained case).The solid line denotes the difference in elastic free energy between single- and double-stranded state. The dotted line denotes the duplex stability free energy at 20 C. The dashed line is the sum of duplex stability and entropic penalty $\Delta S = -2.06 \, k_B$ due to the constraint (KP calculation). Free energies are in units of $k_B T$. The intersection of dashed snd full line defines the overstretch threshold at 111 pN, in agreement with experiment [van Mameren *et al.* (2009)]. (From [Theodorakopoulos (2019)]).

Chapter 7

Helix-coil theory of DNA melting

7.1 Statistical models of helix-coil equilibria

DNA melting curves can be interpreted using a mesoscopic-level modeling. At this level, one can describe a base-pair as being either in a bound, double-helical, or in an unbound, random coil-like state. This reduced description enabled, during the 1960s, the development of theories based on the language and methodology of the Ising model of cooperative phenomena in magnetic and other systems. Much of our current understanding of phase transitions in natural and synthetic biopolymers originates in this type of modeling. In this chapter I will sketch the main physical ideas and techniques; the interested reader is referred to the specialized literature [Poland and Scheraga (1970)] for details.

7.1.1 *Synthetic polypeptides*

Synthetic polypeptides are macromolecules consisting of identical amino-acid residues. They were used extensively in studying the transition from the alpha-helical structure, characteristic of portions of many native proteins, to a denatured, coil-like state. A proper understanding of this helix-coil transition is central to controlling the stability of secondary protein structure (*e.g.*[Creighton (1992)]).

The melting profile of a synthetic polypeptide is similar to that of DNA (Fig. 5.1). The underlying situation is however somewhat simpler. Since there is no strand dissociation, the helix-coil equilibrium

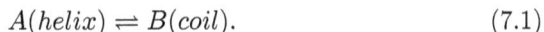

$$A(helix) \rightleftharpoons B(coil). \qquad (7.1)$$

refers to different states of the same molecule. If there are no intermediates ("all or none" conditions), the helix-coil equilibrium constant is independent of the polymer concentration,

$$K \equiv \frac{[B]}{[A]} = \frac{[c(1-\theta)]}{c\theta} \equiv e^{-\Delta G/k_B T} = e^{\Delta S/k_B} e^{-\Delta H/k_B T} \qquad (7.2)$$

and the helix fraction is given by

$$\theta = \frac{1}{1 + e^{-\Delta G/k_B T}}, \qquad (7.3)$$

where $\Delta G, \Delta H, \Delta S$ are, respectively, the free energy, enthalpy and entropy of the transition. As in section 5.1.2.2, a high van't Hoff enthalpy ΔH is a signal of strong cooperativity. Note however that, because of the different dependence of the melting fraction on ΔG, the numerical factor in the equation relating the van't Hoff enthalpy to the slope of the melting curve at midpoint

$$\left(\frac{d\theta}{dT} \right)_{\theta=1/2} = -\frac{\Delta H}{4k_B T_m^2}. \qquad (7.4)$$

is now different.

7.1.2 *Helix growth vs. helix initiation*

Helix *initiation* and helix *growth* are viewed[Zimm and Bragg (1959)] as distinct processes:

- Growth: an existing helix may grow further at the *nth* site, or shrink. This is viewed as a forward and reverse reaction, with a rate ratio

$$s = e^{-\Delta G^*/(k_B T)} \qquad (7.5)$$

  which reflects the difference in local free energies $\Delta G^* = \Delta H^* - k_B T \Delta S^*$ between the helix and coil states. If the ratio is greater than unity, the helix has a tendency to grow ("zip"). If it is less than unity, the helix will shrink ("unzip"). At temperatures near the transition, $s \approx 1$. Depending on the physical context — polypeptide or DNA — the enthalpy difference ΔH^* can be related to the underlying microscopic interactions. In the former case it corresponds to the energy of a single hydrogen bond formed in the process of helix growth. In the latter it represents the sum of hydrogen bond and stacking energies, e.g. as calculated within the nearest-neighbor model (cf. section 5.2.2). In either case helix formation leads to a negative entropy change ΔS^* as previously unconstrained bond angles now acquire specific orientations.

- Nucleation: in order to initiate helix formation, a number of residues — or base pairs, according to the type of biopolymer — have to organize themselves. Again, viewing nucleation as a forward / reverse reaction, introduces a dimensionless $\sigma = \exp(-\Delta G_{init}/(k_B T))$. The large difference in the free energy comes mostly from the entropy loss $\Delta S_{init} < 0$ associated with the organization of the molecular units involved in the first turn of the helix. As a consequence

$$\sigma \approx e^{\Delta S_{init}/k_B} \qquad (7.6)$$

can be regarded as a temperature-independent parameter.

In the next sections I will present some elementary statistical models of the helix-coil transition[Zimm and Bragg (1959)] in order of increasing complexity.

7.1.3 The "all-or-nothing" (AON) model

Only two states are significant within this model. The pure coil, with relative statistical weight equal to unity; and the helix with N residues, with a relative weight σs^N. Intermediate states are suppressed, presumably due to high rate barriers. This gives a partition function

$$Z = 1 + \sigma s^N, \qquad (7.7)$$

a helix fraction

$$\theta = \frac{\sigma s^N}{1 + \sigma s^N} \qquad (7.8)$$

and a slope at midpoint

$$\left(\frac{d\theta}{dT} \right)_{\theta=0.5} = -\frac{N}{4} \frac{\Delta H^*}{k_B T_m^2}. \qquad (7.9)$$

Comparing (7.4) to (7.9) shows that $\Delta H = N\Delta H^*$. The cooperativity of the transition, as measured by the sites involved, is — by construction — complete.

7.1.4 The zipper model

The model allows a single connected helical region of any length n smaller or equal to the total number of sites N. The statistical weight (Boltzmann factor) is -according to the general considerations, cf. above-, σs^n, and the helix can commence at any of the first $A_n = N - n + 1$ positions. This gives a partition function

$$Z = 1 + \sum_{n=1}^{N} A_n \sigma s^n \tag{7.10}$$

and a helical fraction

$$\theta = \frac{1}{Z} \sum_{n=1}^{N} n A_n \sigma s^n = \frac{s}{Z} \frac{\partial Z}{\partial s}, \tag{7.11}$$

where the partition sum can be evaluated to give

$$Z(N) = 1 + \sigma s^{N+2} - (N+1)s + \frac{N}{(s-1)^2}. \tag{7.12}$$

7.1.5 The generalized zipper model

In this case the macromolecule consists of any number of helical and coil regions which may alternate freely. One associates the following statistical weights with the different types of alternation:

- 1 if coil comes after helix or coil;
- s if helix comes after helix;
- σs if helix comes after coil (nucleation).

The model thus implements the ideas presented in section 7.1.2 without imposing any further constraints. The helical or coil-like state of the residue (or base pair) at site i can be described by a 2-vector ν_i, and the partition function is given by

$$
\begin{aligned}
Z_N &= \sum_{\{\nu_1\}...\{\nu_N\}} <\nu_1|T|\nu_2><\nu_2|T|\nu_3> ... <\nu_{N-1}|T|\nu_N> \\
&= \sum_{\{\nu_1\},\{\nu_N\}} <\nu_1|T^N|\nu_N>
\end{aligned} \tag{7.13}
$$

where the matrix elements of $\mathbf{T}$ express the Boltzmann factors specified above, i.e.

$$\mathbf{T} = \begin{pmatrix} s & 1 \\ \sigma s & 1 \end{pmatrix}. \tag{7.14}$$

To facilitate the evaluation of the partition sum, I apply periodic boundary conditions (an artificial condition which does not alter significantly the properties of long chains) and obtain

$$Z_N = Tr\mathbf{T}^N = z_0^N + z_1^N \tag{7.15}$$

where the eigenvalues of the characteristic equation of $\mathbf{T}$

$$(z - s)(z - \sigma s) - 1 = 0 \tag{7.16}$$

are given by

$$z_{0,1} = \frac{1}{2}\left[1 + s \pm \Delta\right], \tag{7.17}$$

$$\Delta = \sqrt{(1 - s)^2 + 4\sigma s}$$

and, in the large N limit, Z is dominated by the largest eigenvalue, z_0.

Note that the partition function (although not the $\mathbf{T}$ -matrix directly!) and the eigenvalues (7.18) can be mapped onto the corresponding quantities of the ferromagnetic Ising model:

$$\mathcal{H} = -J \sum_{i=1}^{N-1} \tau_i \tau_{i+1} - h \sum_{i=1}^{N} \tau_i \tag{7.18}$$

with $\{\tau_i = \pm 1\}$, J the exchange interaction and h the magnetic field. The mapping identifies

$$s \Leftrightarrow e^{-2\beta h}, \tag{7.19}$$

$$\sigma \Leftrightarrow e^{-2\beta J}, \tag{7.20}$$

$$z_{magnetic} = e^{\beta(J+h)} z_{helix-coil}. \tag{7.21}$$

To obtain the helix fraction, note that the probability of obtaining a helical segment of length k is given by $\phi_k(\sigma)ks^k$, where ϕ is the coefficient of s^k in the partition sum. This gives

$$\theta = \frac{1}{N}\frac{1}{Z}\sum_{k=1}^{N}\phi_k(\sigma)ks^k = \frac{1}{N}\frac{s}{Z}\frac{\partial Z}{\partial s} \tag{7.22}$$

It may now be verified that, as $s \to 1$, $\Delta \to 2\sqrt{\sigma}, \theta \to 1/2$; for $\sigma << 1$ (cf. below),

$$\left(\frac{d\theta}{dT}\right)_{\theta=0.5} = \frac{1}{4\sqrt{\sigma}}\frac{\Delta H^*}{T_m^2}. \tag{7.23}$$

Two comments are in order here:

- One usually interprets $1/\sqrt{\sigma}$ as number of residues cooperatively involved in the transition (cf. AON theory, Eq. 7.9). This interpretation also follows from the Ising model, where correlations between distant spins are known to decay exponentially

$$< \tau_i \tau_j > \propto e^{-|i-j|/\xi} \tag{7.24}$$

with a correlation length ξ given (in units of the lattice constant) by

$$\xi = \frac{1}{z_1 - z_0} = \frac{1}{2\Delta} \approx \frac{1}{2\sqrt{\sigma}} \text{ (at } s = 1). \qquad (7.25)$$

- although the helix-coil transition described within the AON or the zipper models may be quite sharp, reflecting a large degree of cooperativity, the mathematical models do not exhibit a strict thermodynamic phase transition. In other words, the partition function, and the associated thermodynamic functions derived from it, do not develop a singularity at the melting temperature as $N \to \infty$. This is not just a property of the one-dimensional Ising model, for which one can compute the partition function exactly. It can be shown (cf. Appendix B) for a wider class of one dimensional systems that a thermodynamic phase transition cannot occur at any finite temperature [Landau and Lifshitz (1980)].

7.2   Melting of infinitely long homogeneous DNA. The Poland–Scheraga model

Poland and Scheraga[Poland and Scheraga (1966)] proposed a simple model of the thermodynamics of DNA melting, based on the concepts of helix initiation and growth (cf. section 7.1.2), while also recognizing that non-helical regions are not simply coil-like linear polymers but should be treated as loops of single stranded DNA (cf. Fig. 7.1) with an associated entropic contribution.

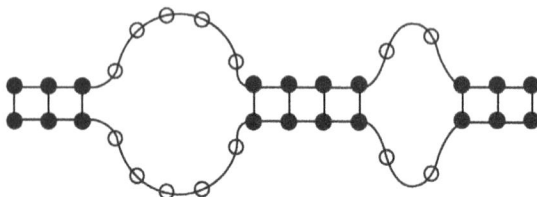

Fig. 7.1   Schematics of the helix-coil picture of DNA. Each helical region is followed by a coil-like region (loop). Filled circles denote bases which are bound in pairs. Open circles denote unpaired bases.

7.2.1 A useful shortcut

It is possible to obtain the thermodynamics of the generalized zipper model without recourse to the transfer matrix formalism. I present this "hand-waving" calculation [Lifson (1964); Azbel (1979)], because it can readily be generalized to include entropic contributions from DNA loops.

The fundamental "entity" of the macromolecule is a helical region of length n, followed by a coil region of length m. I will consider a macro-molecule of infinite length. Each conformation can then be completely characterized by the helical segment initiation sites $\{i_1, i_2, \cdots\}$ and the lengths of successive helical and coil regions $n_{i_1}, m_{i_1}; n_{i_2}, m_{i_2}, \cdots$. Following the ideas developed in 7.1.2 and 7.1.5, I associate a statistical weight σs^n with each fundamental unit of n helical followed by m coil-like sites. Note that the coil sites contribute only factors of unity; statistical weights are defined in such a way as to describe the entropic cost and the energetic advantage of forming a helix (helix nucleation and growth, cf. above), compared to the reference coil-like state.

This "helix-coil" entity is characterized by a free energy

$$g(n, m) = -k_B T \ln \sigma - n k_B T \ln s; \qquad (7.26)$$

it occurs with a probability

$$P_{n,m} = \exp\{-[g(n, m) - (n + m)g_0]/(k_B T)\} \qquad (7.27)$$

where $g_0 \equiv -k_B T \ln z$ is the equilibrium free energy per site of the full macromolecule, to be determined by the normalization condition

$$\sum_{n,m=1}^{\infty} P_{n,m} = 1. \qquad (7.28)$$

Both the n and the m- summations can be performed trivially as long as $s < z$ and $z > 1$. The condition (7.28) can then be written as

$$\sigma \frac{s}{z - s} \frac{1}{z - 1} = 1, \qquad (7.29)$$

whose roots are identical to those of (7.18) obtained via the transfer matrix; the largest root is the one which satisfies the condition $z > \max(1, s)$ (cf. above).

It is now straightforward to calculate the average lengths of helical and coil-like segments. They are given by

$$<n> \equiv \sum_{n,m=1}^{\infty} P(n, m)n = \frac{z}{z - s} \qquad (7.30)$$

and

$$< m > \equiv \sum_{n,m=1}^{\infty} P(n,m)m = \frac{z}{z-1} \qquad (7.31)$$

respectively.

Noting that the sum $< m > + < n >$ expresses the average length of an elementary entity, we arrive at a helical fraction

$$\theta \equiv \frac{< n >}{< m > + < n >} = \frac{z-1}{2z-s-1} = \frac{s}{z}\frac{\partial z}{\partial s} \qquad (7.32)$$

where the latter property can be verified explicitly, making use of the normalization condition (7.29). In fact, this latter expression holds generally true even if we assign purely entropic weights to the coil-like segments, i.e. weights that have no dependence on temperature, and hence no dependence on the parameter s. I will make use of the latter property in the next section.

7.2.2   *Loop entropies*

A denaturation loop of length m, i.e. a region of m sites where the double helix has locally melted, is characterized by the extra entropy it contributes[Poland and Scheraga (1966)]. This extra entropy has been calculated for lattice polymers and is of the form

$$S_L(m) = am + b - c \ln m, \qquad (7.33)$$

where a and b are constants and c depends on the dimensionality and the nature of the polymer which forms the loop. In the case of Gaussian, or freely-jointed polymer chains in three dimensions, $c = 3/2$. Excluded volume effects tend to increase the value of c. It will be seen below that this can have a decisive influence on the existence and/or the nature of the phase transition.

7.2.3   *The phase transition*

In an infinitely long DNA molecule, the fundamental entity is again a double-helical region of n sites (base pairs), followed by a denaturation loop of length m ($2m$ bases); putting together the contributions from helical and loop part I obtain the free energy of this entity as

$$g(n,m) = -k_B T[\ln \sigma + n \ln s - c \ln m] \qquad (7.34)$$

where I have dropped the irrelevant constant b. The constant a can also be set equal to zero, since whatever entropic cost is associated with the helical state is always computed with reference to the coil (loop) state. It is now possible to derive the thermodynamics exactly as in section 7.2.1. Inserting (7.34) in the normalization condition (7.28) gives

$$\frac{z}{s} - 1 = \sigma U_c (1/z),$$ (7.35)

where

$$U_c(x) = \sum_{m=1}^{\infty} \frac{x^m}{m^c} \equiv Li_c(x).$$ (7.36)

is the polylogarithm function and s is the only temperature dependent parameter.

Eq. (7.35) can be numerically solved to provide $z(s)$ and thus a complete description of helix-coil statistics of an infinite DNA homopolymer in the framework of the Poland–Scheraga model. To illustrate exactly what happens it will be necessary to follow its properties in some detail. This necessitates listing some basic properties of the polylogarithm function for the physically relevant real values of its argument $0 < x < 1$ and the parameter $c > 0$.

7.2.3.1 Relevant properties of the polylogarithm function

- the sum defined in (7.36) is convergent for any $|x| < 1$.
- The derivative of the polylogarithm function satisfies

$$Li_c'(x) = \frac{1}{x} Li_{c-1}(x).$$ (7.37)

- For values of the parameter $c > 1$, the polylogarithm function evaluated at $x = 1$ reduces to the Riemann zeta function

$$Li_c(1) = \zeta(c); \quad c > 1.$$ (7.38)

- In the special case $c = 1$, the polylogarithm reduces to the usual logarithmic function

$$Li_1(x) = -\ln(1 - x).$$ (7.39)

- In the special case of $c = 2$ the so-called dilogarithm function satisfies the identity

$$Li_2(x) + Li_2(1 - x) = \zeta(2) - \ln x \ln(1 - x).$$ (7.40)

- For noninteger c and x near 1 the dependence on $\delta = 1 - x$ is non-analytical:

$$Li_c(1-\delta) = \Gamma(1-c)\delta^{c-1}[1+\mathcal{O}(\delta)]+\zeta(c)-\zeta(c-1)\delta+\mathcal{O}(\delta^2)].$$ (7.41)

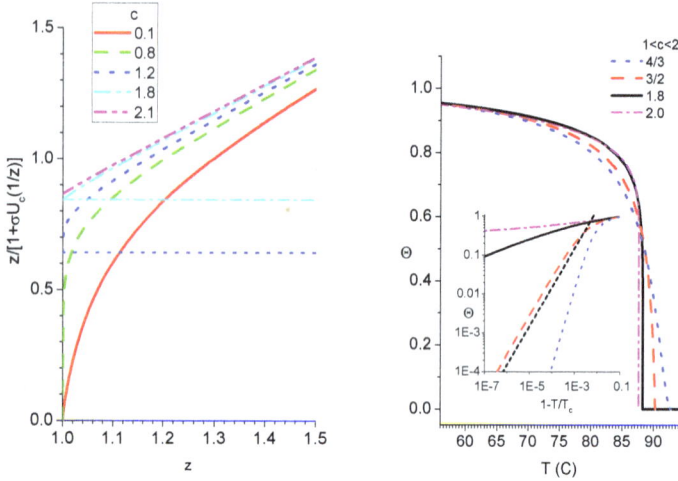

Fig. 7.2   *Left panel:* Graphical solution of (7.35). The function $z/[1 + \sigma U_c(1/z)]$ is plotted as a function of z for various values of c. Note that for the lowest two values of c, there will always exist a root z, no matter how small the value of s. This is not the case if $c > 1$, where roots, as indicated by the intersections of the curves with horizontal lines, exist only if $s > s_c$. The upper boundary of the shaded area is the straight line $y = z$. It is therefore clear that any roots will automatically satisfy the conditions $z > s$, $z > 1$. *Right panel:* The helical fraction as a function of temperature in the case of a continuous transition, i.e. $1 < c \leq 2$. Details of the approach to T_c are shown in the inset on a log-log scale. Note how the slope of the limiting curves, initially larger than unity if $c < 3/2$, decreases to 1 at $c = 3/2$ and decreases further as $c > 2$. In the limiting case $c = 2$ the melting fraction approaches zero as $1/(- \ln(T_c - T))$. An extra short-dashed line of points with slope equal to unity has been added as a guide to the eye.

7.2.3.2   *A phase transition occurs only if $c > 1$.*

We are now equipped to proceed with the analysis of (7.35). For the purposes of graphical analysis it is more convenient to rewrite it as

$$\frac{z}{1 + \sigma Li_c(1/z)} = s. \tag{7.42}$$

The idea is now to find a root $z(s)$ with the properties $z > 1$ and $z > s$ which are necessary in order to ensure the convergence of the summations leading to (7.35). Given $z(s)$ it is then straightforward to compute the helical fraction (cf. Eq. 7.32))

$$\theta = \frac{s}{z} \frac{\partial z}{\partial s} = \frac{1 + \sigma Li_c(1/z)}{1 + \sigma Li_c(1/z) + \sigma Li_{c-1}(1/z)}, \tag{7.43}$$

where z is determined from (7.42), and I have used the property (7.37).

The graphical analysis of the root-finding procedure can be followed at the left panel of Fig. 7.2. For the helix initiation and propagation parameters I have used, respectively, $\sigma = 0.1$, and

$$s = e^{\Delta S^*/k_B} e^{-\Delta H^*/(k_B T)},$$

with $\Delta H^* = -7.9$ kcal/mol and $\Delta S^* = -22.2$ cal/mol. For $c \leq 1$ the limiting value $Li_c(1)$ diverges. A root $z(s)$ of (7.42) will exist for any finite value of s, no matter how small. There is no singularity of z as a function of s, and therefore none as a function of temperature either. The helical fraction remains finite at all temperatures. A thermodynamic phase transition does not occur.

If $c > 1$ the limiting value of $Li_c(1)$ is finite, equal to the value of the Riemann zeta function at c. The graphical analysis now indicates that a root can only be found if s exceeds a critical value s_c

$$s_c = \frac{1}{1 + \sigma \zeta(c)}, \tag{7.44}$$

corresponding to a critical temperature

$$T_c = \frac{\Delta H^*}{\Delta S^* + k_B \ln[1 + \sigma \zeta(c)]}. \tag{7.45}$$

7.2.3.3 *The phase transition is continuous if* $1 < c \leq 2$ ...

Although the *existence* of the phase transition only demands that $c > 1$, its *detailed character* depends further on the exact value of c. According to (7.43) the value of the helical fraction near T_c depends on both Li_c and Li_{c-1}. For $1 < c < 2$, the expansion (7.41) of L_{c-1} for $z = 1 + \delta$, $\delta \ll 1$, is dominated by the divergent term

$$Li_{c-1}\left(\frac{1}{1+\delta}\right) \sim \Gamma(2 - c)\delta^{c-2}, \tag{7.46}$$

which allows me to rewrite (7.35) to leading order in δ and $s - s_c$ as

$$\frac{s - s_c}{s_c^2} \sim -\sigma \Gamma(1 - c)\delta^{c-1}. \tag{7.47}$$

Inserting (7.46) in (7.43) and making use of (7.47) finally leads to

$$\theta \propto (T_c - T)^{\frac{2-c}{c-1}}. \tag{7.48}$$

The helical fraction approaches zero continuously as $T \to T_c$. Nonetheless, the detailed shape of the melting curve itself (cf. the right panel of Fig.7.2) still depends crucially on the value of c. For $c < 3/2$, the exponent in (7.48) exceeds unity and the helical fraction approaches zero with a horizontal tangent near T_c. Conversely, if $c > 3/2$, the exponent is smaller than unity and helical fraction approaches zero at T_c with an infinite slope. The marginal case $c = 3/2$ - which physically corresponds to the loop entropy of a random, three-dimensional Gaussian chain and was the case originally considered [Poland and Scheraga (1966)]-, leads to an exponent equal to 1 and a linear slope of the melting curve at T_c. The case $c = 1.8$ corresponds very closely to the self-avoiding random walk in three dimensions[Fisher (1966a)].[1]

7.2.3.4   ... and discontinuous if $c > 2$.

If $c > 2$, Eqs. (7.44) and (7.45 still hold. Moreover, both polylogarithmic functions in (7.43) now approach finite values; To leading order now

$$\frac{s_c - s}{s_c^2} \approx -\sigma \zeta(c-1)\delta, \tag{7.49}$$

and, as temperature approaches the melting temperature from below, (7.43) gives a well defined limit

$$\lim_{T \to T_c^-} \theta = \theta_c = \frac{1 + \sigma\zeta(c)}{1 + \sigma\zeta(c) + \sigma\zeta(c-1)}. \tag{7.50}$$

In other words, there is now a finite discontinuity of the melting fraction at T_c. The details of the melting curve are shown in Fig. 7.3.

The above analysis shows how crucial the value of c is in determining the mathematical nature of the DNA melting transition within the context of the Poland–Scheraga model. Subsequent research[Kafri *et al.* (2000)] suggested that a full account of excluded volume effects may lead to a corrected value of c as high as 2.1. The origin of the correction lies in the fact that looping occurs not at the ends but at interior sites of the polymer chain. If the distance between successive loops is sufficiently large, the corrected values of the exponent c, which are always derived in the asymptotic limit, may indeed apply.

It should however be noted that, from an experimental point of view, analyzing data with a finite temperature resolution, there is not much difference between a continuous melting profile where the melting fraction approaches zero with a vertical tangent (cf. right panel of Fig. 7.2) and

---

[1]The exponent for the SAW is 1.75. Note that this is really the exponent $-\beta$ of 1.5.2, characterizing the number of walks returning to the origin; the loop entropy (7.33) is the logarithm of that number.

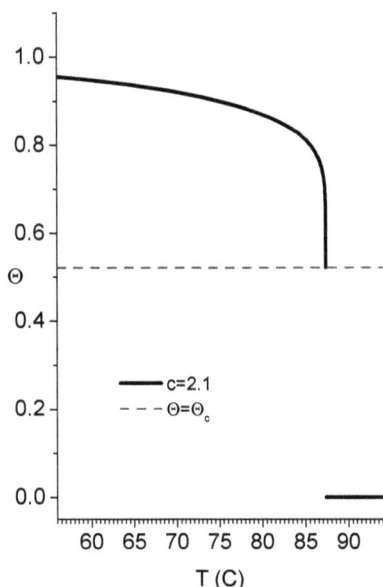

Fig. 7.3   The helical fraction as a function of temperature in the case of a discontinuous transition, i.e. $c = 2.1$. Note that, immediately prior to the discontinuity there is a steep drop of the helical fraction.

a discontinuous one (cf. Fig. 7.3). Moreover,the description of multistep melting occurring as a consequence of sequence heterogeneities, does not demand an exact discontinuous transition in the underlying homogeneous system. In practical terms this is very well confirmed by a standard program which successfully computes detailed DNA melting profiles [Blake et al. (1999)] using $c = 1.7$.

7.2.4   Bubbles and clusters

In some applications, it is useful to have an estimate of the average size of intact helical regions (in brief: clusters) ξ_c, or of locally denatured regions (loops, also known as denaturation bubbles) ξ_b. Both can be calculated, making use of the definitions (7.30), (7.31), respectively, and of the constitutive thermodynamic equation (7.35). The results,

$$\xi_c = \frac{z(s)}{z(s) - s} \tag{7.51}$$

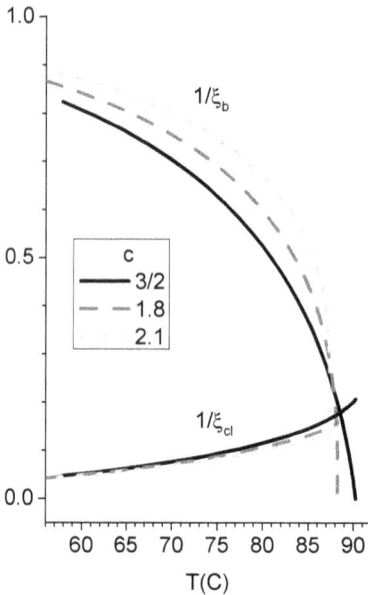

Fig. 7.4   The inverse of the average size (in monomer distance units) of bubbles (upper set of curves) and clusters (lower set of curves) for a range of values of the loop entropy parameter. Note the discontinuity which occurs for $c > 2$ at the critical temperature.

and

$$\xi_b = \frac{Li_{c-1}[1/z(s)]}{Li_c([1/z(s)]}$$

(7.52)

respectively, are plotted for the parameter set described in section 7.2.3.2, in Fig. 7.4. Again it should be noted that, as long as the transition is continuous, the bubble size diverges at the critical point. For a first order transition however this is no more the case. A bubble can grow to encompass a few base pairs on average and then suddenly the thermodynamic phase transition occurs. Note however that, at temperatures far from melting, the average bubble size is always close to unity, independently of the transition's order.

7.2.5   *A   thermodynamic   phase   transition   in   a   one-dimensional system?*

Thermodynamic transitions, i.e. macroscopic transformations of the equilibrium phase occurring as a result of a variation of temperature, are exceptional in one-dimensional systems and usually reflect some type of pathology in the underlying interactions. Various theorems prove the non-existence of such transitions under certain conditions which exclude and/or limit pathological interaction features. I will return to this point in Appendix B. Regarding however the PS model, the situation is quite straightforward. The chain composed of alternating helical and coil-like regions is not really one-dimensional. Its defining feature is a macroscopic number of loops embedded in 3-dimensional space, with entropic contributions controlled by the embedding space.

Chapter 8

Dynamical theory of DNA melting I. Fundamentals

8.1 Base pairs as dynamical entities

8.1.1 *Functional requirements*

The double helix is far from forming a rigid structure under normal physiological conditions. Storage of the genetic code would be meaningless without an efficient mechanism for its retrieval. The *transcription* process which produces RNA copies of individual genes starts with a local, enzyme-controlled opening of the double helix.

8.1.2 *Breathing of individual base pairs*

Spontaneous local conformational fluctuations, also known as "breathing" of individual DNA base pairs, were observed long ago by von Hippel and coworkers[Printz and von Hippel (1965); von Hippel and Wong (1971)]. These thermal fluctuations, which occur below the melting temperature, lead to a local, temporary breakup of the hydrogen bonds in a base pair and expose the bases to the surrounding solvent. Using hydrogen isotope exchange methods, it became possible to follow the rate at which hydrogen binding the duplex is exchanged with that of the solvent and thus demonstrate the "breathing" of base pairs.

Not all hydrogen atoms belonging to the DNA duplex are exchangeable with those of the surrounding water. Some are strongly bonded to nitrogen (aliphatic hydrogens). Some of those which are exchangeable (exterior amino hydrogens) are characterized by very fast relaxation times. The relevant information for base-pair opening is contained in the exchange rates of interior amino and imino hydrogens, both of which are involved in base pairing. GC pairs have two of the former and one of the latter, AT pairs have one of each. Thus, there is one imino proton per base pair.

8.1.3 *Base pair lifetime*

Gueron and coworkers [Gueron *et al.* (1987)] examined the NMR spectra of imino protons in oligonucleotides. Their work suggests that imino proton exchange is a two step process: first, the base pair must open, and then, while in the intermediate open state, the imino proton of the base pair is exchanged with that of the solvent. In the absence of appropriate catalytic agents, the exchange process is slow, and this is what the NMR relaxation rates measure. By introducing an appropriate catalyst and measuring relaxation rates as a function of its concentration, they were able to extrapolate to the limit of infinite catalyst concentration and determine the base pair lifetime, as well as the equilibrium constant of the open-closed reaction. At room temperatures the ratio of open to closed pairs was determined to be of the order of 10^{-5} for AT and 10^{-6} for GC pairs; the base pair lifetime is of the order of 10 ms.

8.1.4 *Low frequency vibrations*

Several low-frequency vibrational modes have been identified in various DNA conformations [Weidlich *et al.* (1990)]. Longitudinal phonons were first detected in B-DNA fibers by means of inelastic light scattering [Maret *et al.* (1979)]. The measured sound velocity in dry fibers was 3800 m/sec; more recent inelastic neutron scattering results [van Eijck *et al.* (2011)] suggest a slightly higher value.

Of direct relevance to the denaturation transition is perhaps a Raman active mode at a frequency of 85 cm^{-1} which has been detected[Urabe and Tominaga (1981)] in an aqueous solution of DNA. This particular mode was identified as transverse and was found to become "soft", i.e. its frequency decreases as the temperature rises, ultimately disappearing from the spectrum at temperatures higher than 80 C.

Prohofsky and coworkers [Gao *et al.* (1984)] have identified large amplitude phonon motion associated with the hydrogen bonds which bind DNA base pairs. These phonon modes cluster in a frequency range between 70 and 110 cm^{-1} and would account for the spectroscopic observation in random-sequence DNA. Modeling the hydrogen bond anharmonic motion in terms of Morse potentials and taking account of thermal motion in an average, yet self-consistent fashion (self-consistent phonon theory) leads to a prediction of "hydrogen bond melting" associated with phonon mode softening at a temperature of 70 C [Gao *et al.* (1984)].

8.2 The homogeneous Peyrard–Bishop model

8.2.1 *Definitions and Notation*

Peyrard and Bishop (PB) [Peyrard and Bishop (1989)] proposed a minimal, mesoscopic model of DNA melting which incorporates much of the discussion of the preceding section. The model involves a single relevant degree of freedom per base pair, describing the displacement of the two bases relative to each other and transverse to the helical axis. Under normal conditions, this relative displacement performs small oscillations around its primary equilibrium state. As temperature increases however, each base pair begins to explore an alternative metastable equilibrium state. This is modeled by a nonlinear local potential with a plateau extending to infinity. In addition, in order to account for the tendency of neighboring base pairs to stack, the model includes a nonlocal, harmonic coupling between nearest-neighbor base pairs.

The above model assumptions result in a Hamiltonian of the form

$$H = \sum_{j=1}^{N} \frac{p_j^2}{2\mu} + \frac{1}{2}k \sum_{j=1}^{N-1} (y_{j+1} - y_j)^2 + \sum_{j=1}^{N} V(y_j), \tag{8.1}$$

where the transverse coordinate y_j represents the separation of the two bases at the jth site, p_j the canonically conjugate momentum, μ is the reduced mass of a base pair, and k is a measure of the tendency of neighboring pairs to stack;[1] finally, the local term, an on-site Morse potential, depicted in Fig. 8.2, with depth D and range $1/\alpha$,

$$V(y) = D(1 - e^{-\alpha y})^2 \tag{8.2}$$

describes the combined effects of hydrogen-bonding, stacking and solvent acting on each base pair.

In the following, I will denote the total potential energy, i.e. the sum of the second and third terms in (8.1) by H_P.

---

[1]This is not the same as "the strength of the stacking interaction", since the latter can only be given if the interaction is explicitly attractive, i.e. a potential well. In this case, however, the stacking interaction is modeled as a repulsive interaction. Proper stacking corresponds to the "least repulsive" configuration, i.e. when two neighboring base pairs are equally displaced from their stable equilibrium positions and the repulsive energy vanishes.

8.2.2 *Dynamics*

The Hamiltonian equations of motion are

$$\dot{y}_j = \frac{\partial H}{\partial p_j} = \frac{1}{\mu} p_j$$

$$\dot{p}_j = -\frac{\partial H}{\partial y_j} = k[y_{j+1} + y_{j-1} - 2y_j] - \frac{\partial V(y_j)}{\partial y_j}, \tag{8.3}$$

or, eliminating the momenta,

$$\mu \ddot{y}_j = k[y_{j+1} + y_{j-1} - 2y_j] - V'(y_j), \tag{8.4}$$

where the dots represent time derivatives[2].

8.2.2.1 *Small oscillations*

If the displacements are small, (8.4) reduces to a set of linear, coupled, second order differential equations

$$\mu \ddot{y}_j = k[y_{j+1} + y_{j-1} - 2y_j] - 2D\alpha^2 y_j. \tag{8.5}$$

Under fixed boundary conditions (cf. footnote 2), (8.5) admits solutions of the type

$$y_j^{(q)}(t) \propto \sin(qaj)\sin(\Omega_q t) \tag{8.6}$$

("normal modes"), where a is the distance between nearest neighbor sites, q the wavevector, which must take any of the discrete values

$$q_\nu = \frac{\pi \nu}{(N+1)a}, \quad \nu = 1, \cdots, N,$$

in order to satisfy the boundary conditions, and Ω_q the frequency of oscillation. Frequency and wavevector are related by the *phonon dispersion relation*

$$\Omega_q^2 = 4\omega_0^2 \sin^2\left(\frac{qa}{2}\right) + \Omega_0^2, \tag{8.7}$$

where $\omega_0^2 = k/\mu$ and $\Omega_0^2 = 2D\alpha^2/\mu$. The phonon dispersion relationship is illustrated in Fig. 8.1 and is characterized by the existence of a *gap*. There are no frequencies lower than Ω_0. This type of oscillation is known in solid-state physics as an "optical phonon".

---

[2]In deriving the above dynamical equations I have, for the sake of notational compactness, neglected the effect of boundary sites. Equations (8.3) and (8.3) retain their validity at the boundaries if we extend the system by two further sites, at $j = 0$ and $j = N + 1$, with displacements y_0 and y_{N+1} held fixed at zero ("fixed boundary conditions").

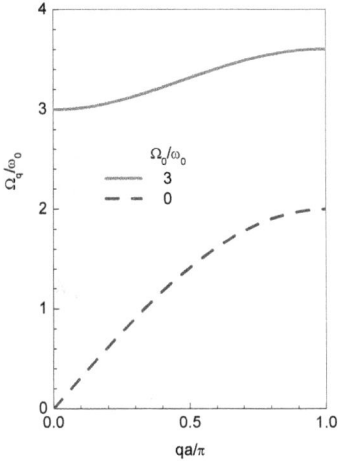

Fig. 8.1   The phonon dispersion relationship (8.7) illustrated in the case $\Omega_0/\omega_0 = 3$ (optical phonon, solid line) and in the "soft" limit $\Omega_0/\omega_0 = 0$ (acoustic phonon, dashed line).

8.2.3   Statistical mechanics

8.2.3.1   *The partition function*

The classical thermodynamics of (8.1) can be derived [Dauxois *et al.* (2002)] from the canonical partition function

$$Z = \int_{-\infty}^{\infty} dp_1 \int_{-\infty}^{\infty} dy_1 \cdots \int_{-\infty}^{\infty} dp_N \int_{-\infty}^{\infty} dy_N e^{-\beta H}. \tag{8.8}$$

One can immediately perform the Gaussian integrals over momentum space and obtain

$$Z = Z_K Z_P, \tag{8.9}$$

where each integration in the kinetic part contributes a $(2\pi\mu/\beta)^{1/2}$ factor to the partition function, i.e.,

$$Z_K = (2\pi\mu/\beta)^{N/2}, \tag{8.10}$$

and the configurational part of the partition function is given by

$$Z_P = \int_{-\infty}^{\infty} dy_1 \cdots \int_{-\infty}^{\infty} dy_N e^{-\beta H_P(y_1,\cdots,y_n)} \tag{8.11}$$

$$= \int_{-\infty}^{\infty} dy_1 \cdots \int_{-\infty}^{\infty} dy_N \, e^{-\beta V(y_1)/2} \, T(y_1, y_2)$$

$$T(y_2, y_3) \cdots T(y_{N-1}, y_N) \, e^{-\beta V(y_N)/2}, \tag{8.12}$$

with

$$T(x,y) = e^{-\beta/2\left[\mu\omega_0^2(y-x)^2+V(x)+V(y)\right]}.\tag{8.13}$$

There are no generally applicable, exact methods to compute the classical configurational partition function of a one-dimensional continuous system. An important general method, valid when displacements are small and the harmonic approximation can be justified, consists of transforming (8.11) to a product of Gaussian integrals. Another useful, exact approach for homogeneous systems is to exploit the factorization in (8.12). Both methods have the advantage that they generate numerically computable schemes for *inhomogeneous* systems and are therefore useful in the context of this book.

8.2.3.2   *The Gaussian approximation*

If displacements are small, as is the case e.g. at sufficiently low temperatures, we may use a low-order expansion of the configurational part of the Hamiltonian around its global minimum at $\{y_k = 0 \quad \forall k\}$. Keeping second-order terms in the displacements results in

$$H_P = \frac{1}{2}k\sum_{j=1}^{N-1}(y_{j+1}-y_j)^2 + D\alpha^2\sum_{j=1}^{N}y_j^2.\tag{8.14}$$

(8.14) is a special case of a quadratic form

$$H_P = H_P(\{y_j = 0\}) + \frac{1}{2}\sum_{i,j=1}^{N}M_{ij}y_iy_j,\tag{8.15}$$

where the second derivative matrix $\mathbf{M}$, with elements

$$M_{ij} = \left.\frac{\partial^2 H_P}{\partial y_i\partial y_j}\right|_{\{y_k=0\forall k\}},\tag{8.16}$$

is known as the Hessian. The partition function (8.11) can be now approximated as

$$Z_P = \int_{-\infty}^{\infty}dy_1\cdots\int_{-\infty}^{\infty}dy_N e^{-\frac{1}{2}\beta\mathbf{y}^T\mathbf{My}}\tag{8.17}$$

where I have adopted a more compact vector notation $\mathbf{y} \equiv (y_1,\cdots,y_N)$ and $\mathbf{y}^T$ is the transpose of $\mathbf{y}$. Now, since $\mathbf{M}$ is real, symmetric, and by virtue of (8.14) positive definite, it can be diagonalized by an orthogonal transformation $\mathbf{D} = \mathbf{T}^T \mathbf{M} \mathbf{T}$. Transforming to a new set of coordinates $\mathbf{z} = \mathbf{T}^T \mathbf{y}$ leads to

$$Z_P = \prod_{\nu=1}^{N} \int_{-\infty}^{\infty} dz_i e^{-\frac{1}{2}\beta\lambda_\nu z_\nu^2}$$

$$= \prod_{\nu=1}^{N} \left(\frac{2\pi}{\beta\lambda_\nu} \right)^{1/2} \tag{8.18}$$

where λ_ν is the ν-th eigenvalue of $\mathbf{M}$ and I have made use of the fact that the Jacobian of the coordinate transformation is unity.

Eq. (8.18) can be applied to any system where anharmonicities are small, e.g. at sufficiently low temperatures. In the case of the PB Hamiltonian under fixed boundary conditions, we recognize that $\Lambda_\nu = \mu\Omega_{q_\nu}^2$, and therefore, combining with (8.10), we obtain

$$Z = \prod_{\nu=1}^{N} \left(\frac{2\pi}{\beta\Omega_{q_\nu}} \right). \tag{8.19}$$

Eq. (8.19) expresses a general relationship between the low-temperature classical partition function and the normal mode frequencies of any system.

8.2.3.3 *The transfer integral (TI)*

Consider the eigenvalue problem

$$\int_{-\infty}^{\infty} dy\, T(x,y)\, \phi_\nu(y) = \Lambda_\nu \phi_\nu(x) \tag{8.20}$$

associated with the symmetric kernel $T(x,y)$. If it can be shown to have a complete

$$\sum_\nu \phi_\nu(y)\phi_\nu(y') = \delta(y' - y) \tag{8.21}$$

set of orthonormal

$$\int_{-\infty}^{\infty} dy\, \phi_\nu(y)\phi_{\nu'}(y) = \delta_{\nu\nu'} \tag{8.22}$$

eigenstates with a nondegenerate eigenvalue spectrum bounded from above, i.e.,

$$\Lambda_0 > \Lambda_1 > \cdots, \tag{8.23}$$

then the kernel can be expanded in terms of its eigenfunctions

$$T(y_i, y_j) = \sum_\nu \Lambda_\nu \phi_\nu(y_i)\phi_\nu(y_j). \tag{8.24}$$

Inserting (8.24) in (8.12) $N - 1$ times, and performing $N - 2$ internal integrations by invoking the orthonormality condition (8.22) — a procedure known as the transfer integral (TI) technique[Scalapino *et al.* (1972)] — I obtain

$$Z_P = \sum_\nu \Lambda_\nu^{N-1} I_\nu^2, \tag{8.25}$$

where

$$I_\nu = \int_{-\infty}^{\infty} dy \, e^{-\beta V(y)/2} \phi_\nu(y) \tag{8.26}$$

are the contributions from the two end integrals.

In the thermodynamic limit $N \to \infty$. the sum (8.25) is dominated by the largest eigenvalue, i.e.,

$$\lim_{N \to \infty} \frac{1}{N} \ln Z_P = \Lambda_0.$$

Computing the thermodynamic functions of the infinitely long, homogeneous PB chain is therefore reduced to finding the largest eigenvalue of (8.20).

8.2.3.4 *The order parameter. Correlations*

Using similar transfer integral techniques, one can express other thermodynamic averages in terms of the spectral properties of (8.20). For example the average displacement at any site of the infinite, homogeneous chain

$$< y_i > = \int_{-\infty}^{\infty} dy \, |\phi_0(y)|^2 \, y \quad \forall i, \tag{8.27}$$

which I will later identify as the order parameter of the system, can be expressed in terms of the eigenfunction associated with the largest eigenvalue.

Correlations of the displacement at different sites

$$< \delta y_i \delta y_{i+r} > \equiv < y_i y_{i+r} > - < y_i >< y_{i+r} >$$
$$= \left| \int_{-\infty}^{\infty} dy \, \phi_0(y) y \phi_1(y) \right|^2 \left(\frac{\Lambda_1}{\Lambda_0} \right)^r \tag{8.28}$$

are essentially controlled by the ratio between the two largest eigenvalues.

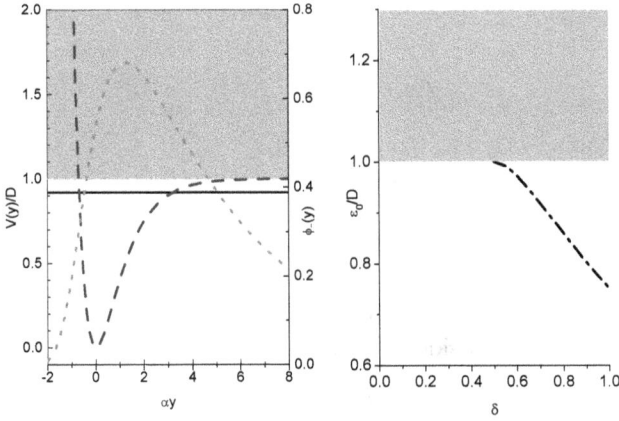

Fig. 8.2 *Left panel:* The Morse potential (8.2) (dashed line). The eigenvalue spectrum of (8.35) for $\delta = 0.7$ consists of a single bound state at $E = 0.9183D$ and a continuum of scattering states for $E > 1$. Also shown (dotted line, right-y axis) is the eigenfunction corresponding to that single bound state. *Right panel:* Merging of the single remaining bound state with the continuum in the limit $\delta \to 1/2$.

8.2.3.5 *The gradient expansion*

Is it possible to transform the integral eigenvalue problem (8.20) to a - presumably more manageable - equivalent differential equation? The answer is yes, under certain conditions to be specified below.

I set

$$\Lambda_\nu \equiv e^{-\beta \epsilon_\nu} \tag{8.29}$$

and

$$\tilde{\phi}_\nu(y) \equiv e^{-\beta V(y)/2} \phi_\nu(x)$$

and rewrite (8.20) as

$$\int_{-\infty}^{\infty} dy \, e^{-\frac{1}{2}\beta \mu \omega_0^2 (y-x)^2} \tilde{\phi}_\nu(y) = e^{\beta[V(x)-\epsilon_\nu]} \tilde{\phi}_\nu(x), \tag{8.30}$$

Suppose now that displacements of neighboring base pairs should not differ substantially from each other. Note that this is not the same as demanding small nonlinearities. It is a weaker statement which enables us to substitute individual displacements by a continuum displacement field and use a gradient expansion for the eigenfunctions

$$\tilde{\phi}_\nu(x+z) = \tilde{\phi}_\nu(x) + \tilde{\phi}'_\nu(x)z + \frac{1}{2}\tilde{\phi}''_\nu(x)z^2 + \cdots. \tag{8.31}$$

Such a continuum approximation will be justified as long as the width of the Gaussian is smaller than the typical scale of variation of the local potential, i.e.

$$\beta\mu\omega_0^2/\alpha^2 > 1, \tag{8.32}$$

which will hold at sufficiently low temperatures.

Performing the Gaussian integrals in the left hand side I obtain

$$\tilde{\phi}_\nu(x) + \frac{1}{2}\frac{1}{\beta\mu\omega_0^2}\tilde{\phi}_\nu''(x) + \cdots = \left(\frac{\beta\mu\omega_0^2}{2\pi}\right)^{1/2} e^{\beta[V(x)-\epsilon_\nu]}\tilde{\phi}_\nu(x), \tag{8.33}$$

where higher order derivatives have been neglected. I can now (i) absorb the prefactor in the right hand side by defining

$$\tilde{\epsilon}_\nu = \epsilon_\nu - \frac{1}{2\beta}\ln\left(\frac{\beta\mu\omega_0^2}{2\pi}\right), \tag{8.34}$$

(ii) divide both sides by $\tilde{\phi}_\nu(x)$, (iii) take the logarithm of both sides, and (iv) multiply again both sides by $\tilde{\phi}_\nu(x)$. Keeping the leading-order term in step (iii) leads to

$$-\frac{1}{2\mu(\beta\omega_0)^2}\tilde{\phi}_\nu''(x) + [V(x) - \tilde{\epsilon}_\nu]\tilde{\phi}_\nu(x) = 0 \tag{8.35}$$

which is a Schrödinger-like equation. With the correspondence

$$1/\beta = k_B T \longleftrightarrow \hbar\omega_0$$

we recognize it as the quantum mechanical equation describing the motion of a particle of mass μ in a Morse potential.[3]

Collecting results from (8.9) and dropping the tildes for notational clarity, I can write for the free energy per base pair

$$g = -\lim_{N\to\infty}\frac{1}{\beta N}\ln(Z_K Z_P) = -\frac{1}{\beta}\ln\left(\frac{2\pi}{\beta\omega_0}\right) + \epsilon_0, \tag{8.36}$$

where the first term can be identified with the free energy of harmonic oscillations (phonons) around the minimum of the Morse potential well, and the second term ϵ_0, the *lowest* eigenvalue of (8.35), originates in the nonlinear, large-amplitude motion.

---

[3]This is another instance of the deeper mathematical analogy between the classical statistical mechanics of a one-dimensional system and the quantum mechanics of a zero-dimensional one (more generally, d-dimensional classical statistical mechanics with $(d-1)$−dimensional quantum mechanics; in section 2.3.2 a similar analogy was established between the statistical mechanics of the Heisenberg model, the behavior of the Kratky–Porod chain under the action of an external force, and the quantum mechanics of the isotropic rotator in an electric field.

8.2.4 The phase transition

8.2.4.1 A singularity in the eigenvalue spectrum...

The eigenvalue spectrum of (8.35) [Morse (1929); Landau and Lifshitz (1977)] is nondegenerate. It consists of a number of bound states

$$\frac{\epsilon_n}{D} = 1 - \left[1 - \frac{n + 1/2}{\delta}\right]^2$$
$$n = 0, 1, ..., int(\delta - 1/2), \tag{8.37}$$

where $\delta = (2\mu D)^{1/2}\beta\omega_0/\alpha$ is a dimensionless parameter, and a continuum of scattering states with $1 < \epsilon/D < \infty$.

There is at least one bound state if $\delta > 1/2$. For $1 \geq \delta > 1/2$ there is *exactly* one bound state. As δ approaches $1/2$, the single remaining bound state approaches the continuum of scattering states (cf. Fig. 8.2), ultimately merging with it at $\delta = 1/2$. For values $\delta < 1/2$ there are no bound states. The value $\delta_c = 1/2$ is therefore "critical". In quantum mechanical language, if a particle has a mass which is lighter than a critical mass $\mu_c = \hbar^2\alpha^2/(8D)$, it cannot be confined in the Morse well. Quantum fluctuations will drive it out[4]. In the context of DNA statistical mechanics, $\delta_c = 1/2$ corresponds to a critical temperature

$$T_c = \frac{2}{k_B} \cdot \frac{\omega_0}{\alpha}\sqrt{2\mu D} = \frac{2}{k_B\alpha}\sqrt{2kD} \tag{8.38}$$

at which the two strands unbind.

8.2.4.2 ...leads to second-order phase transition...

From (8.36) and (8.37) I obtain the free energy (leaving out the nonsingular contribution of small amplitude, harmonic phonons) as a function of temperature,

$$\frac{g}{D} = \begin{cases} 1 & T > T_c \\ 1 - \left(1 - \frac{T}{T_c}\right)^2 & T < T_c, \end{cases} \tag{8.39}$$

where in the upper line I have made use of the fact that the bottom of the continuum part of the spectrum is at $\epsilon = D$. The free energy g is

---

[4]This is a general property of asymmetric one-dimensional wells; symmetric wells will support a particle in a bound state, no matter how low its mass.

non-analytic at $T = T_c$. Its first derivative, the entropy,

$$s = -\frac{\partial g}{\partial T} = \begin{cases} 0 & T > T_c \\ -\frac{2D}{T_c}\left(1 - \frac{T}{T_c}\right) & T < T_c, \end{cases} \tag{8.40}$$

is continuous at the critical point (cf. (8.3)), whereas the second derivative is discontinuous, i.e. there is a jump in the specific heat[5].

8.2.4.3   *. . . with a diverging order parameter. . .*

In order to gain some further insight into the physics involved it is useful to examine the average displacement (8.27), determined by the ground-state (GS) eigenfunction

$$\phi_0(x) = \sqrt{\frac{\alpha}{\Gamma(2\delta - 1)}} e^{-\zeta/2} \zeta^{\delta - 1/2} \tag{8.41}$$

where $\zeta = 2\delta e^{-\alpha x}$ and Γ denotes the gamma function. It is straightforward to see that, as T approaches T_c from below, the eigenfunction extends towards larger and larger positive values of x with a limiting exponential form:

$$\phi_0(x) \propto e^{-x/\xi_\perp} \tag{8.42}$$

where

$$\xi_\perp = \frac{1}{2\alpha} \cdot \frac{1}{\delta - \delta_c} \tag{8.43}$$

is a (transverse) characteristic length which measures the spatial extent of the bound state eigenfunction. As a consequence, we can estimate that $< y >$, which is dominated by the large values of the argument, will also behave as

$$< y >= \int_{-\infty}^{\infty} dy\, y\, \phi_0^2(y) \sim \xi_\perp \propto (\delta - \delta_c)^{-1} \propto \left(1 - \frac{T}{T_c}\right)^{-1}. \tag{8.44}$$

For comparison, I quote the exact result ([Dauxois *et al.* (2002)])

$$< y >= \frac{1}{\alpha} \left[\ln(2\delta) - \psi(2\delta - 1)\right], \tag{8.45}$$

---

[5]This corresponds to a second order transition, according to the Ehrenfest classification scheme of thermodynamic phase transitions. This means that "second order" is meant literally in this case, not just as a loose metaphor for the absence of a latent heat (for which the term "continuous transition" would be appropriate).

where the asymptotic behavior (8.44) can be derived directly from the properties of the ψ — also known as the digamma — function [Abramowitz and Stegun (1964)].

As the critical temperature is approached from below, particles cease to be confined to the minimum of the Morse well. They perform larger and larger excursions to the flatter part of the potential. At T_c the transition is complete; the average transverse displacement is infinite. Particles move, on the average, on the flat top of the Morse potential. Unwinding ("melting") of the DNA has occurred.

In the language of critical phenomena $< y >$ is the order parameter. In "usual" thermodynamic phase transitions, raising the temperature leads from an ordered to a disordered phase. The order parameter vanishes at the transition point. In the PB model however, the competing metastable equilibrium being at infinity, we are dealing with a thermodynamic instability[6], rather than with an "order-disorder" transition. The order parameter diverges.

Experimental data on DNA denaturation do not deliver $< y >$ directly. The "experimental order parameter" is the helical fraction (7.32), i.e. the probability that a given base pair is still bound; in the context of the PB model one uses an (instrumentation-dependent) cutoff y_c and computes the probability that the bases of any given base pair are no more than a distance y_c apart,

$$P_i(y < y_c, T) = \lim_{N \to \infty} \frac{1}{Z_P} \int_{-\infty}^{\infty} dy_1 \cdots \int_{-\infty}^{y_c} dy_i \int_{-\infty}^{\infty} dy_N \, e^{-\beta V(y_1)/2}$$

$$T(y_1, y_2) \, T(y_2, y_3) \cdots T(y_{N-1}, y_N) \, e^{-\beta V(y_N)/2}$$

$$\approx \int_{-\infty}^{y_c} dy \, \phi_0^2(y), \tag{8.46}$$

which, in the limit $N \to \infty$, is independent of i and can be compared with the measured helical fraction. Using (8.41) one can rewrite [Dauxois *et al.* (2002)]

$$P(y < y_c, T) = 1 - \frac{\gamma(2\delta - 1, 2\delta e^{-\alpha y_c})}{\Gamma(2\delta - 1)} \tag{8.47}$$

where γ is the incomplete gamma function.

For the model presented here, this function approaches zero smoothly (linearly) as $T \to T_c$, independently of the choice of y_0. This is exactly the behavior found in the Poland–Scheraga case with the Gaussian chain loop entropy exponent $c = 3/2$ (cf. section 7.2.3.3).

---

[6]A related instability is the wetting of interfaces, where many of the ideas discussed here have been developed[Kroll and Lipowsky (1983); Lipowsky (1985)]

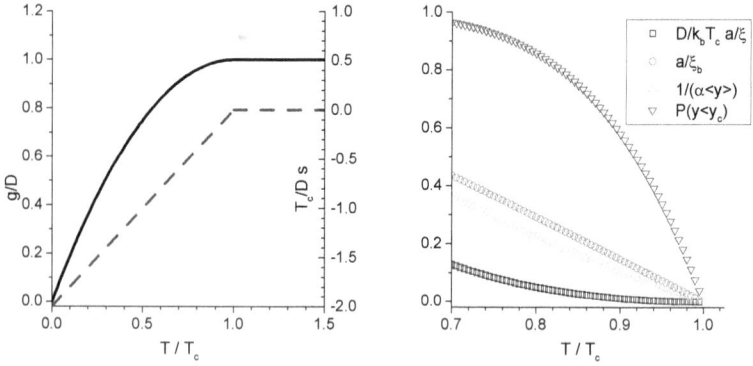

Fig. 8.3   *Left panel:* Dependence of PB model thermodynamic functions on the reduced temperature T/T_c. The free energy (8.39) in units of D (solid line, left y-axis) and the entropy (8.40) in units of D/T_c (dashed line, right y-axis). *right panel:* Critical quantities vs. reduced temperature T/T_c for the PB model: squares, the inverse correlation length (8.48); circles, the inverse of the average bubble size; uptriangles, the inverse of the average transverse displacement, (8.45); down triangles, the helical fraction, (8.47), computed for $y_c\alpha = 8.4$.

8.2.4.4   *... and a diverging correlation length.*

Using (8.28) and (8.29) we can rewrite correlations of displacements at different sites as

$$< \delta y_i \delta y_{i+r} > \propto e^{-\beta(\epsilon_1 - \epsilon_0)r}$$

which identifies the characteristic correlation length, measured in units of the monomer distance a, as

$$\frac{\xi}{a} = \frac{k_B T}{\epsilon_1 - \epsilon_0} = \frac{k_B T}{D}\left(1 - \frac{T}{T_c}\right)^{-2}. \qquad (8.48)$$

A divergent correlation length is a universal feature of continuous thermodynamic phase transitions. At the critical point itself, where ξ becomes infinite, no characteristic length scale exists. Thermodynamic singularities of this type can be described in general terms, even in models which are not exactly solvable, by the renormalization group theory.

8.2.5   *Bubbles*

It is tempting to associate the correlation length with the average spatial extent of a denaturation bubble, where the latter is understood as a contiguous set of sites with a large average transverse displacement, i.e. typically larger than the denaturation cutoff y_c which was introduced previously in the context of the experimental order parameter. Such an association would be in accordance with standard concepts of the theory of critical phenomena [Kadanoff *et al.* (1967); Fisher (1967)] according to which the correlation length is a measure of the largest region in which a coherent fluctuation of the order parameter (i.e. a "bubble" of the "wrong" phase) can occur.

A connection — although not quite an identity — between the correlation length and the average bubble size can indeed be formulated in the PB model as well. It has been shown [Theodorakopoulos (2008)] that the probabilities $\{P_n\}$ for the occurrence of a size-n bubble have the property

$$\frac{P_n}{P_{n-1}} = e^{-1/\xi}\left(1 - \frac{c}{n} + \cdots\right), \quad n \gg 1, \tag{8.49}$$

where c is a weakly temperature-dependent quantity approaching $3/2$ as $T \to T_c$. Consequently, the distribution of bubble sizes in the PB model near the critical point is well approximated by

$$P_n = A\frac{1}{n^{3/2}} \cdot e^{-n/\xi}, \tag{8.50}$$

where A is some normalization constant. The above property allows us to interpret the inverse of the correlation length, i.e. the gap between the two lowest eigenvalues of the Schrödinger-like equation (8.35), as the *activation free energy necessary for the growth of the bubble by one site*. Near the critical point this activation energy tends to zero, and bubbles can grow to very large sizes. *If* one could neglect the second term in parentheses in (8.49), ξ would be equal to the average bubble size. In fact, the correction term is essential in determining critical behavior:

$$\xi_b \equiv \frac{\sum_{n=1}^{\infty} nP_n}{\sum_{n=1}^{\infty} P_n} = \frac{Li_{1/2}(e^{-1/\xi})}{Li_{3/2}(e^{-1/\xi})}, \tag{8.51}$$

where $Li(x)$ is the polylogarithm function introduced in 7.2.3. From the known asymptotics of the polylogarithm function it is possible to estimate that

$$\xi_b \propto \xi^{1/2} \propto (T_c - T)^{-1},$$

i.e. the average bubble size exhibits a weaker, linear divergence near T_c — in contrast to the quadratically divergent ξ. The exact quantity (8.51) is plotted in Fig. 8.3.

8.2.6 *DNA unzipping in the PB model*

A transverse external force f acting on the first site adds a term

$$H_f = -f y_0$$

to the PB Hamiltonian (8.1). Considering again that the partition function (8.25) is dominated by the largest eigenvalue Λ_0, we have to examine the modified end term in (8.26) which now becomes

$$I_0(f) = \int_{-\infty}^{\infty} dy \, e^{\beta f y} \phi_0(y).$$

Noting that the eigenfunction decays exponentially with a characteristic length $\xi_\perp$ we conclude that I_0 can only remain finite as long as

$$f < f_c \equiv \frac{k_B T}{\xi_\perp}; \tag{8.52}$$

in other words the PB-modeled DNA has a thermodynamic instability ("unzips") at a critical force f_c. Moreover, since $\xi_\perp \propto (T_c - T)$, the $f - T$ phase diagram is such that the boundary between bound and unbound states is a straight line which approaches the temperature axis linearly, with a finite slope (cf. Fig. 6.2 where the boundary approaches the temperature axis with an infinite slope). Again, this is a reflection of the PB model property of a second-order thermodynamic phase transition.

Chapter 9

Dynamical theory of DNA melting II. Nonlinear stacking interaction

9.1 The homogeneous Peyrard–Bishop–Dauxois (PBD) Hamiltonian

The dynamical model of DNA melting presented in the previous chapter should really be viewed as a limited proof of principle. It demonstrates that a true, second-order thermodynamic instability can indeed arise as thermal fluctuations *cooperatively* drive locally oscillating entities to the flat part of their local potential well. Important salient features of the observed melting transition are however missing. We recall from the thermodynamic analysis of Chapter 5:

- Transition enthalpies are characteristic of neighboring base pair dimers (rather than single base pairs); therefore, any modeling of the underlying base stacking interactions has to reflect this pair property.
- The observed melting transition is — in contrast to what the PB model predicts — effectively first order.

Both aspects have been successfully incorporated in the Peyrard–Bishop–Dauxois (PBD) Hamiltonian [Dauxois *et al.* (1993)], which modifies the harmonic nearest-neighbor interaction term in (8.1) to include a "nonlinear base stacking" contribution,

$$\frac{1}{2}k[1 + \rho e^{-b(y_{j+1}+y_j)}](y_{j+1} - y_j)^2, \tag{9.1}$$

where $1/b$ represents the range of the nonlinear stacking interaction and ρ its strength relative to the linear term.

The interaction energy (9.1) interpolates between the original (residual) value of the elastic coupling k if *either* (or both) of the two base pairs $j, j-1$

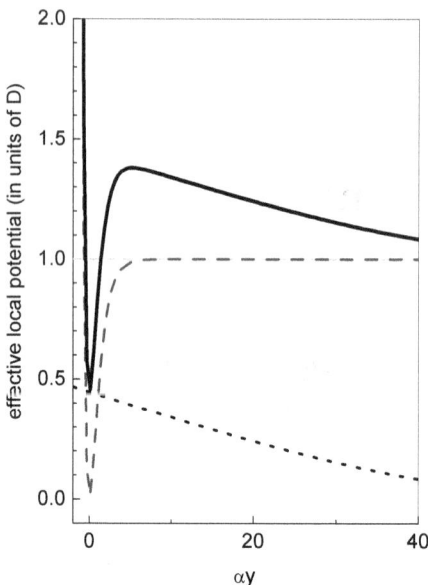

Fig. 9.1   The combined effect of the Morse potential (dashed curve) and the thermal
barrier (9.2) generated by the nonlinear stacking interaction (dotted curve) is represented
by the solid curve. The parameters used here are $D = 0.1255$ eV, $T = 330K$, $b/\alpha =$
0.0476, $\rho = 50$. Because of the much longer range of the stacking interaction, compared
with that of the Morse potential, critical behavior will be dominated by the former.

moves out of the stack (in which case $y_j \to \infty$ or $y_{j-1} \to \infty$, respectively),
and the full value $k(1 + \rho)$ if both are bound; in the latter case, typically,
$y_j \approx 0$; note that the much higher values which (9.1) in principle allows
for negative y_j's, are statistically irrelevant due to the repulsive core of the
Morse potential.

I will discuss the choice of correct PBD model parameters separately,
in the context of long, heterogenous chain melting. However, it is worth
noting at this stage that at least some choices are mandated by geometry and flexibility. For example, the radius of the double helix, $10.5A$,
defines the order of magnitude, and sets an upper limit on the range of
the stacking interaction. The choice of $b = .2A^{-1}$, made in the present
chapter, corresponds to $1/b = 5A$ and is therefore consistent with what we
know about double-helical geometry. Furthermore, the interpolation logic
of the nonlinear stacking interaction, should at least qualitatively reflect

the large difference between double-stranded (ds-) and single-stranded (ss-) DNA bending stiffness — the former being mainly related to the presence of stacking forces. The value of $\rho = 50$ chosen to characterize the relative strengths of restoring forces for stacked vs. unstacked neighboring base pairs is roughly consistent with the ratio of observed room temperature ds- and ss-DNA persistence lengths.

9.2   A local thermally activated barrier

Within the gradient expansion approximation, the main effect of the nonlinear stacking energy on the thermodynamics is to generate an effective, on-site, "thermally activated" barrier [Cule and Hwa (1997); Theodorakopoulos *et al.* (2000)]. This can be readily seen by inspection of (8.30). Neglecting higher order terms in the gradient expansion, the effect of introducing the interaction (9.1) is equivalent to multiplying the mass by a factor $(1 + \rho e^{-2bx})$. This has the effect of modifying the subsequent integration by the square root of that factor, or, equivalently, introducing an extra term in the local potential energy,

$$U(x) = \frac{1}{2} k_B T \ln \left(1 + \rho e^{-2bx} \right), \qquad (9.2)$$

which appears in Eq. 8.35 and acts in addition to the Morse potential.

It has been shown that the character of the transition changes dramatically as the value of the stacking parameter ratio b/a decreases (corresponding to a longer range of the effective thermal barrier vs. that of the Morse potential, cf. Fig. 9.1). Details of how this happens will be presented in the next section. At this point I will simply summarize the upshot of the gradient-expansion-based analysis, according to which, as long as $b/a \ll 1$, although the transition remains asymptotically second order, the limiting asymptotic behavior becomes relevant only within an exponentially vanishing interval of the temperature difference $T_c - T$. Therefore, for all practical purposes, the transition is first order with a finite melting entropy $\Delta S = A_0 D/T_c$, where A_0 is a numerical constant of order unity[Theodorakopoulos *et al.* (2000)].

9.3   Thermodynamic properties of the PBD Hamiltonian

In the presence of the thermal barrier it is no more possible to find an exact solution of the Schrödinger-like equation. Rather than pursue an approximation scheme to solve what is already an approximation (since it has been obtained by use of the gradient expansion) it is possible to look for a numerical solution of the original integral eigenvalue problem

$$\int_{-\infty}^{\infty} dy\, T(x,y)\, \phi_\nu(y) = \Lambda_\nu \phi_\nu(x) \tag{9.3}$$

with the symmetrized kernel appropriate to the PBD model

$$T(x,y) = e^{-\beta W(x,y)} e^{-\beta[V(x)+V(y)]/2} \tag{9.4}$$

where

$$W(x,y) = \frac{1}{2}k[1 + \rho e^{-b(x+y)}](y - x)^2 \tag{9.5}$$

and V is the Morse potential (8.2).

Before proceeding with the numerical solution scheme, it should be recognized that, owing to the repulsive core of the Morse potential, the lower limit of (9.3) at minus infinity is redundant. The kernel becomes negligible at a value y_{min} equal to minus a few multiples of $1/\alpha$, the range of the Morse potential. Furthermore, the upper limit will be approximated by some finite cutoff L, typically 50-100 times the range of the stacking interaction. A proper control of the numerical procedure involves checking that the results do not depend on the choice of cutoffs. In particular any systematic dependence on the upper cutoff L should disappear in the limit $L \to \infty$. I will return to that point in the next section.

It is now possible to approximate the integral eigenvalue equation (9.3) defined in the finite interval (y_{min}, L), by a matrix eigenvalue problem, using e.g. Gauss-Legendre quadratures [Abramowitz and Stegun (1964)]. Details of how to do this are provided in Appendix D.

Fig. 9.2 presents three typical eigenvalue spectra of (9.3), numerically determined using Gauss-Legendre quadratures with a mesh of 1201 points in the interval $(-1.5A, 300A)$. The model parameters referring to the Morse potential, $\alpha = 4.2A^{-1}$, $D = 0.1255eV$, and the harmonic part of the stacking interaction, $k = .45meV/A^2$, are common to all three spectra. The left panel shows the case of zero nonlinear stacking for reference. The middle and right panel illustrate the case of strongly nonlinear stacking with $\rho = 50$ and, respectively, $b = 0.2\ A^{-1}$ (long range stacking force) and $b = 2\ A^{-1}$ (short range stacking force). The difference between the two last panels is striking. In the first case, where the range of the stacking force has been chosen according to the geometrical constraints imposed by the double helix, the discrete eigenvalue appears to cut into the continuum. Not so when the stacking force is of short range (right panel); in that case the spectrum qualitatively resembles that of the left panel (PB model, harmonic stacking force) where the discrete eigenvalue approaches the continuum in a smooth fashion (quadratically). In other words, the long-range nature of the nonlinear base-stacking forces as modeled in the PBD Hamiltonian would seem to be responsible for a change of the order of the thermodynamic transition from continuous (second-order) to discontinuous (first-order).

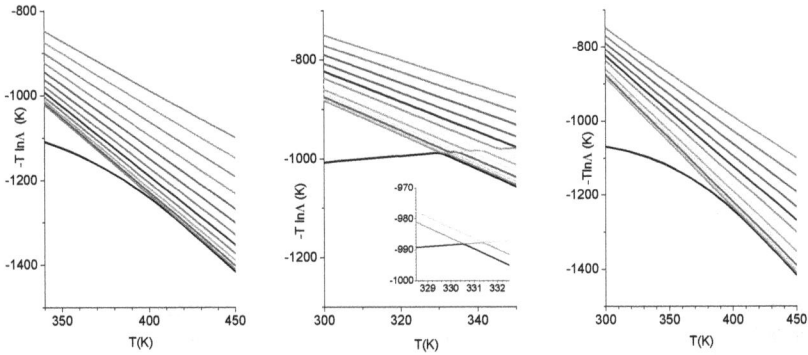

Fig. 9.2   *Left panel:* Temperature dependence of the numerically determined eigenvalue spectrum of the TI operator (8.20) for the PB Hamiltonian. The first 12 eigenvalues are shown, starting with the highest Λ_ν, corresponding to the lowest ϵ_ν (cf. Eq. 8.29). Note the smooth, quadratic approach of the discrete eigenvalue toward merging with the continuum. *Middle panel:* same for the PBD Hamiltonian, i.e. including the nonlinear stacking interaction (9.1) with a ratio of range parameters $b/\alpha = 0.0476$ (cf. Fig. 9.1). Note how the discrete eigenvalue appears to "cut" into the continuum. The inset, containing points with a resolution higher than $.1K$, shows no signs of an avoided crossing between the first and the second eigenvalues, as mandated by the nondegeneracy of the spectrum. The gap between the discrete state and the continuum effectively vanishes linearly with temperature. *Right panel:* same as in the middle panel except for $b/\alpha = 0.476$, i.e. the base-stacking interaction is now of short range, comparable to that of the Morse potential. Note that the spectrum now resembles that of the PB Hamiltonian (left panel).

The detailed thermodynamic properties of the PBD model with a long-range nonlinear stacking interaction also follow what is expected of a first-order transition. For example, the helical fraction drops very sharply from unity to zero within a temperature range of less than a tenth of a degree (Fig. 9.3, left panel). The same figure reveals a concomitant steplike rise in the entropy per base pair which amounts to approximately $4k_B$. Finally, the gap separating the discrete eigenvalue of the TI matrix from the bottom of the continuum band drops linearly to zero with a rounding visible at a few hundredths of a degree, before it starts rising again.

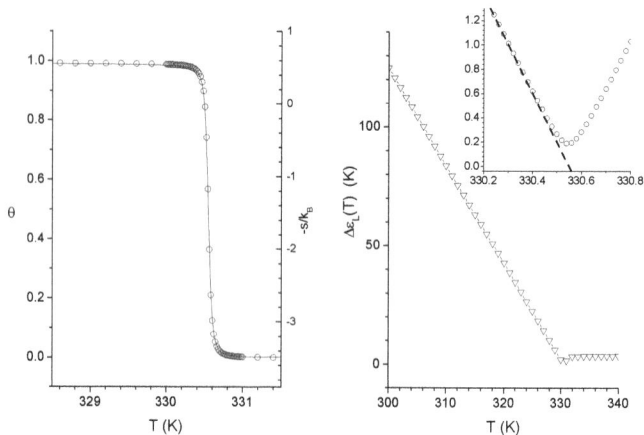

Fig. 9.3   *Left panel:* Open circles, left y-scale: Temperature dependence of the numerically determined helical fraction for (8.20) for the PBD Hamiltonian. Parameters are the same as used in the middle panel of Fig. 9.2. Solid curve, right y-scale: (Minus) the entropy per base pair of the PBD model as a function of temperature, in units of the Boltzmann constant. Note that the (negative of) the entropy is almost exactly proportional to the helical fraction. *Right panel:* Temperature dependence of the gap between the two highest eigenvalues of the PBD Hamiltonian. Parameters are the same as used in the middle panel of Fig. 9.2. The inset zooms in the immediate vicinity of the transition point. Note the essentially linear approach towards zero, which persists down to few hundredths of a degree.

9.4   First-order transition: true or apparent?

We noted already in passing that the approximation of the TI integral equation by a matrix eigenvalue problem results in finite rounding effects, mostly owing to the finite upper cutoff L. A proper check of the numerical procedure always involves examining the extent to which results are cutoff-dependent. For example, if we want to examine whether the gap approaches zero asymptotically, we can systematically increase L while at the same time refining the integration mesh (in order to keep numerical quality constant).

At each value of L, the temperature at which the gap in the spectrum attains its minimum value is taken to be the finite-size approximation $T_m(L)$ to the critical temperature. We can then follow the dependence of the gap on the reduced temperature $t = 1 - T/T_m(L)$, as usually done in the theory of critical phenomena. The result for the PBD model with long-range nonlinear stacking interaction ($b = 0.2A^{-1}$) is shown in the left panel of Fig. 9.4. As the value of L increases, rounding is pushed to lower values of the dimensionless temperature t. At the highest L considered, the gap vanishes linearly with temperature down to $t = 10^{-5}$. This would seem to provide a convincing numerical demonstration of a first-order transition. And yet, caution is required. The eigenvalue spectra obtained for a higher value of b (cf. Fig. 9.2a right panel) clearly suggest a smoother (apparently second-order) vanishing of the gap as the transition is approached. Accordingly, it is just possible that we are dealing with a "crossover" scenario. In other words, the transition could in principle always second-order but, sufficiently far from the transition point, i.e. as long as $T < T_{cr}$, or $|t| > t_{cr}$, this behavior is masked by transient terms which favor a first-order transition.

In order to explore this scenario, it is useful to repeat the $L-$dependent analysis for a slightly higher value, e.g. $b = 0.33\overset{\circ}{A}^{-1}$, which however still qualifies as "long-range"$(b/\alpha = 0.0786 \ll 1)$. The results, shown in the

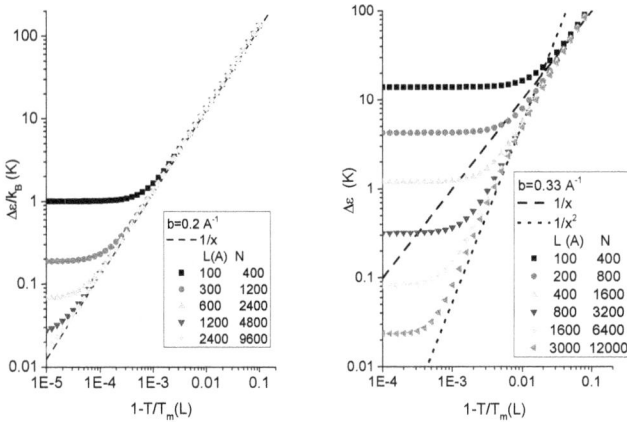

Fig. 9.4   *Left panel:* Temperature dependence of the numerically determined gap in the TI spectrum of the PBD Hamiltonian for various values of the integration cutoff L. Also shown are the numbers of the corresponding integration mesh points. Parameters are the same as used in the middle panel of Fig. 9.2. As the value of L increases, rounding is pushed to lower values of the dimensionless temperature t. At the highest L considered, the gap vanishes linearly with temperature down to $t = 10^{-5}$. *Right panel:* Same for a slightly higher value of $b = .33$. Note the bending of the curves which originally start off linearly but then approach quadratic behavior at $|t| < 0.02$. The dashed and dotted lines provide guides to the eye for linear and quadratic behavior respectively.

right panel of Fig. 9.4 confirm exactly this hypothesis. At the lowest values $L = 100$ the gap appears to follow approximately linear behavior; note however that the temperature range covered before numerical rounding sets in is confined to $|t| > 0.02$. Numerical data obtained for higher values of L, while confirming this approximately linear behavior for $|t| > 0.02$ clearly reveal a bending at temperatures closer to the critical point. Data obtained at lower values of $|t|$ appear to converge toward quadratic behavior.

In summary, the critical behavior of the PBD model with a long-range stacking interaction, as revealed by a systematic study of size-dependent numerics, follows a crossover scenario. Asymptotic behavior is always second-order, but this - depending on the range of the stacking interaction - is only revealed at temperatures very close to T_c. As a consequence, for realistic DNA parameter values, where t_{cr} is lower than 10^{-5}, corresponding to a few thousands of a degree in the temperature difference $T_c - T_{cr}$, the *observed* transition will always be dominated by transient first-order behavior.

Chapter 10

Dynamical Theory of DNA melting III: long, heterogeneous chains

10.1 Heterogeneity in the PBD model framework

10.1.1 *Theory describes* internal *melting*

In the two previous chapters it has been demonstrated that a "minimal" one-dimensional model, motivated by nonlinear lattice dynamics, can in principle describe the onset of a macroscopic thermodynamic instability in a homogeneous chain. The real test of the model's relevance to DNA is its potential ability to describe melting in long, heterogeneous chains. This means taking sufficient account of sequence effects by introducing a manageable parametrization and testing whether the model can actually predict melting profiles for a given sequence and salinity. Why long, rather than short, or intermediate chains are decisive has been explained in Section 5.3. In essence, if one wants to analyze *purely* sequence effects, i.e. what has been defined as "internal melting", ideally one would like to examine profiles where the effects of "external melting" are negligible. This will always be the case if chains are sufficiently long. In practice this means $N > 500$.

10.1.2 *The Hamiltonian and its model parameters*

In general, the configurational part of the PBD Hamiltonian will contain site-dependent contributions from both types of interaction, local and nonlocal. In the last chapter it has been made abundantly clear that the nonlinear contributions to the nonlocal interaction are essential in describing the stacking contribution. Nonetheless, in the interest of computational and notational simplicity, here I will restrict the discussion of heterogeneity effects to the local interaction. This is the equivalent of taking account of "site-disorder" while neglecting "bond-disorder" in discussing the effects of randomness in solid-state theory. It results in the potential energy function

$$H_P = \sum_{j=1}^{N-1} W(y_j, y_{j+1}) + \sum_{j=1}^{N} V_j(y_j) , \qquad (10.1)$$

where

$$W(y_j, y_{j+1}) = \frac{k}{2}[1 + \rho e^{-b(y_j + y_{j+1})}](y_j - y_{j+1})^2 \qquad (10.2)$$

describes the nonlinear base-stacking interaction without specific reference to sequence effects, and

$$V_j(y_j) = D_j(1 - e^{-\alpha_j y_j})^2 \qquad (10.3)$$

extends the PBD Hamiltonian to take account of local heterogeneity effects. The values of the parameters k, ρ, b adopted in this chapter are therefore those of section 9.3. The values of α_j will be taken as 4.2 Å^{-1} or 6.9 Å^{-1}, depending on whether the jth site refers, respectively, to an *AT* or *GC* pair. Finally the values of D_j will depend not only on the type of base pair but, in addition, on the salinity of the solution (cf. below, section 10.4.1).

10.2   Statistical mechanics of the finite, heterogeneous chain

10.2.1   *The partition function*

The configurational partition function of a heterogeneous chain with N base pairs and open ends,

$$Z_N = \int_{-\infty}^{\infty} dy_1 \cdots \int_{-\infty}^{y_c} dy_n \cdots \int_{-\infty}^{\infty} dy_N \; e^{-\beta H_P}$$

$$= \int_{-\infty}^{\infty} dy_1 \int_{-\infty}^{\infty} dy_2 \cdots \int_{-\infty}^{\infty} dy_N \; e^{-\beta V_1(y_1)/2} \, T_{12}(y_1, y_2)$$

$$T_{23}(y_2, y_3) \cdots T_{N-1 N}(y_{N-1}, y_N) \, e^{-\beta V_N(y_N)/2}, \qquad (10.4)$$

can be computed numerically, in the absence of translational invariance. as an N-dimensional integral. Note that, as in the homogeneous case, the limits at minus and plus infinity can be substituted by lower and upper cutoffs y_{min} and L respectively. Unfortunately, this method can only work for sequences of up to about 100 base pairs. Longer chains not only require excessive computing times, but make it impossible to maintain numerical accuracy.

It is however possible to convert the multidimensional real-space integral (10.4) to a matrix product in the space of eigenfunctions of a "reference kernel" of the homogeneous problem. We choose the homogeneous chain consisting of AT pairs as the reference state. Fig 10.1 shows why the choice of GC would lead to divergent results. The expansion of the reference kernel in terms of its eigenvalues and eigenfunctions (cf (8.24)),

$$T_{ij}^{AT}(y_j, y_{j+1}) = \sum_{\nu_j} \Lambda_{\nu_j} \phi_{\nu_j}(y_j)\phi_{\nu_j}(y_{j+1}), \qquad (10.5)$$

allows us to compute all integrals successively. Using the auxiliary quantities

$$A_\nu^{(j)} = \left(\frac{\Lambda_\nu}{\Lambda_0}\right)^{1/2} \int_{-\infty}^{\infty} dy\, \phi_\nu(y)e^{-\frac{\beta}{2}V_{AT}(y)-\beta\Delta V_j(y)} \quad j=1,N,$$

$$B_{\nu\nu'}^{(j)} = \left(\frac{\Lambda_\nu\Lambda_{\nu'}}{\Lambda_0^2}\right)^{1/2} \int_{-\infty}^{\infty} dy\, \phi_\nu(y)\phi_{\nu'}(y)e^{-\beta\Delta V_j(y)} \quad j\neq 1,N, \quad (10.6)$$

where $\Delta V_j(y) = V_{GC}(y) - V_{AT}(y)$ if j is a GC site and 0 if j is an AT site,

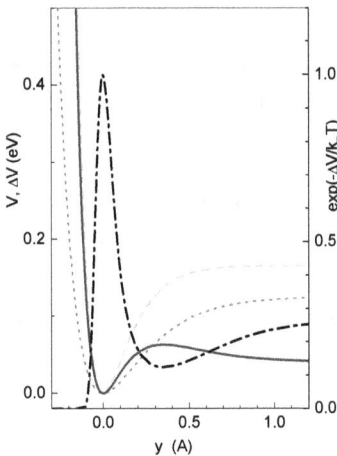

Fig. 10.1   The Morse potential with typical AT parameters, $D = 0.1255eV$, $\alpha = 4.2\text{Å}^{-1}$ (dotted line), GC parameters, $D = 0.1655\, eV$, $\alpha = 6.9\text{Å}^{-1}$ (dashed line), the difference $\Delta V = V_{GC} - V_{AT}$ (solid line). Also shown (dash-dotted line, right y-scale) is the quantity $\exp(-\beta\Delta V)$. Note how the latter quantity vanishes already at slightly negative values of y, while approaching a constant at plus infinity.

it is straightforward to rewrite the partition function as

$$Z_N = \Lambda_0^{N-1} \sum_{\nu_1,\nu_2\cdots\nu_{N-1}} A_{\nu_1}^{(1)} B_{\nu_1\nu_2}^{(2)} B_{\nu_2\nu_3}^{(2)} \cdots B_{\nu_{N-2}\nu_{N-1}}^{(N-1)} A_{\nu_{N-1}}^{(N)} \qquad (10.7)$$

or, in more compact matrix notation,

$$Z_N = \Lambda_0^{N-1} < \mathbf{A}^{(1)} | \mathbf{B}^{(2)} \cdots \mathbf{B}^{(N-1)} | \mathbf{A}^{(N)} > . \tag{10.8}$$

In practice, we will drop terms in the kernel expansion (10.5) which correspond to eigenstates with $\Lambda_\nu / \Lambda_0 < 10^{-8}$. This will restrict the dimensions of the matrix $\mathbf{B}$ and the vector $\mathbf{A}$ to be typically of the order of 100, i.e far smaller than the size of mesh necessary for direct numerical integration in real space according to (10.4).

10.2.2 *The melting fraction*

The probability that the nth base pair is bound is given by

$$p_n = \frac{1}{Z_N} \int_{-\infty}^{\infty} dy_1 \cdots \int_{-\infty}^{y_c} dy_n \cdots \int_{-\infty}^{\infty} dy_N \, e^{-\beta H_P}, \tag{10.9}$$

where y_c is an appropriately chosen crossover distance which distinguishes the open from the bound state of a base pair. I will use the value $y_c = 2\text{\AA}$. However, it should be noted that complex melting profiles exhibit very little sensitivity to the exact choice of y_c.

It is possible to compute the N-dimensional integral in (10.9) in completely analogous fashion with the partition function, as in the previous section. The only difference is that we now need modified auxiliary vectors and matrices, i.e.

$$\mathcal{A}_\nu^{(j)} = \left(\frac{\Lambda_\nu}{\Lambda_0} \right)^{1/2} \int_{-\infty}^{y_c} dy \, \phi_\nu(y) e^{-\frac{\beta}{2} V_{AT}(y) - \beta \Delta V_j(y)}, j = 1, N,$$

$$\mathcal{B}_{\nu\nu'}^{(j)} = \left(\frac{\Lambda_\nu \Lambda_{\nu'}}{\Lambda_0^2} \right)^{1/2} \int_{-\infty}^{y_c} dy \, \phi_\nu(y) \phi_{\nu'}(y) e^{-\beta \Delta V_j(y)}, j \neq 1, N, \tag{10.10}$$

in terms of which

$$p_1 = \frac{< \mathcal{A}^{(1)} | \mathbf{B}^{(2)} \cdots \mathbf{B}^{(N-1)} | \mathbf{A}^{(N)} >}{< \mathbf{A}^{(1)} | \mathbf{B}^{(2)} \cdots \mathbf{B}^{(N-1)} | \mathbf{A}^{(N)} >},$$

$$p_n = \frac{< \mathbf{A}^{(1)} | \mathbf{B}^{(2)} \cdots \mathcal{B}^{(n)} \cdots \mathbf{B}^{(N-1)} | \mathbf{A}^{(N)} >}{< \mathbf{A}^{(1)} | \mathbf{B}^{(2)} \cdots \mathbf{B}^{(N-1)} | \mathbf{A}^{(N)} >}, n = 2, \cdots, N - 1,$$

$$p_N = \frac{< \mathbf{A}^{(1)} | \mathbf{B}^{(2)} \cdots \mathbf{B}^{(N-1)} | \mathcal{A}^{(N)} >}{< \mathbf{A}^{(1)} | \mathbf{B}^{(2)} \cdots \mathbf{B}^{(N-1)} | \mathbf{A}^{(N)} >}. \tag{10.11}$$

The fraction of open pairs, commonly known as the melting fraction, is then given by

$$\theta = 1 - \frac{1}{N} \sum_{n=1}^{N} p_n. \tag{10.12}$$

10.2.3 Computational issues

The astute reader will have noticed that, although substantial computational efficiency has been achieved through the mapping of real-space integration to matrix multiplication in the restricted space of eigenvectors of the homogeneous AT transfer-integral kernel, two computational issues remain open. First, in computing all the p_n's, a lot of matrix multiplication steps must be unnecessarily repeated. In other words, computing the full profile for a chain of N base pairs involves $\mathcal{O}(N^2)$ computational steps. Second, maintaining numerical accuracy in multiplying matrices and vectors over and over can only be achieved if the norms of the intermediate results remain close to unity.

Both of these issues can be addressed by defining appropriate unit vectors at the ends via

$$|\mathbf{v}^{(1)}> \ = \ \frac{1}{\mu_1^R}|\mathbf{A}^{(N)}>, \tag{10.13}$$

$$<\mathbf{u}^{(1)}| \ = \ <\mathbf{A}^{(1)}|\frac{1}{\mu_1^L}, \tag{10.14}$$

where the μ's are the norms of the respective unnormalized vectors. I then store the result of each successive matrix-vector multiplication in vector form, as a unit vector and a norm, i.e.

$$\mathbf{B}^{(N-j)}|\mathbf{v}^{(j)}> \ = \ \mu_{j+1}^R|\mathbf{v}^{(j+1)}>, \tag{10.15}$$
$$<\mathbf{u}^{(j)}|\mathbf{B}^{(j+1)} \ = \ <\mathbf{u}^{(j+1)}|\mu_{j+1}^L, \quad j = 1, 2, \cdots, N-2.$$

It follows that (i)

$$Z_N = \Lambda_0^{N-1}\mu_1^L \cdots \mu_n^L \mu_1^R \cdots \mu_{N-n}^R <\mathbf{u}^{(n)}|\mathbf{v}^{(N-n)}> \tag{10.16}$$

for *any* choice of $n = 1, 2, \cdots, N-1$, and (ii)

$$p_n = \frac{1}{\mu_n^L}\frac{<\mathbf{u}^{(n-1)}|\mathcal{B}^{(n)}|\mathbf{v}^{(N-n)}>}{<\mathbf{u}^{(n)}|\mathbf{v}^{(N-n)}>} \quad n = 2, \cdots, N-1,$$

$$p_1 = \frac{1}{\mu_1^L}\frac{<\mathcal{A}^{(1)}|\mathbf{v}^{(N-1)}>}{<\mathbf{u}^{(1)}|\mathbf{v}^{(N-1)}>},$$

$$p_N = \frac{1}{\mu_N^R}\frac{<\mathbf{u}^{(N-1)}|\mathcal{A}^{(N)}>}{<\mathbf{u}^{(N-1)}|\mathbf{v}^{(1)}>}. \tag{10.17}$$

Computing and storing the necessary intermediates for a melting profile of a chain with N base pairs in vector form is now a process with $\mathcal{O}(N)$ computational steps.

10.3   Computed melting profiles

10.3.1   *The T-7 phage*

The techniques developed in the previous sections have been used [Theodor-akopoulos (2010)] to compute melting profiles of large genomic sequences. As an example I show in Fig. 10.2 the melting profile of the T7 phage, a sequence of 39937 base pairs with a 48.4% GC content at $0.0195M$ molar concentration of Na^+. The experimental data from [Lyubchenko *et al.* (1976)] is also included for comparison. The only fit parameters are the depths of the Morse wells, providing values of $D_{AT} = 0.1205eV$ and $D_{GC} = 0.1619eV$ respectively.

In order to relate melting profiles to sequence details, it is customary to draw a melting map, in which each site is characterized by its melting temperature, defined via $p_i(T_m(i)) = 1/2$. The melting map for the T7 phage is shown in the right panel of Fig. 10.2, along with the moving 200-pt average of local GC-content.

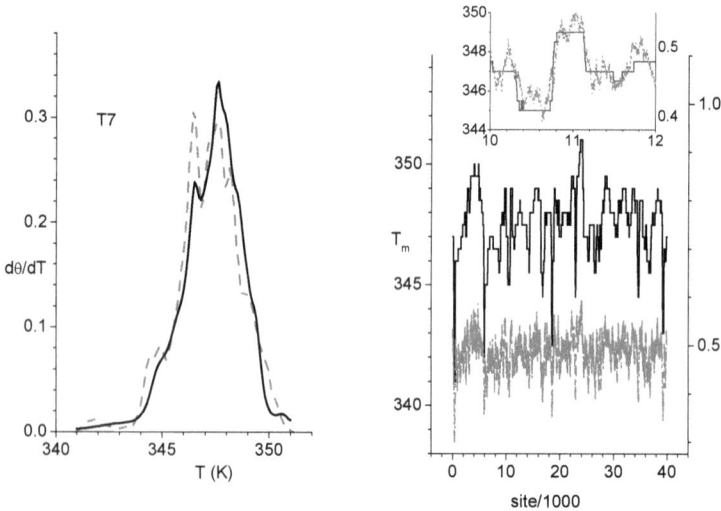

Fig. 10.2   *Left panel:* The differential melting curve of the T7 phage. The full line shows theoretical results based on the PBD model, the dashed curves with the shaded area experimental results redrawn from [Lyubchenko *et al.* (1976)]. *Right panel:* upper curve, the melting map (left y-scale); lower curve, a 200-site moving average of GC content (right y-scale). The inset shows a zoomed region of the melting map with the GC-content curve shifted for clarity. (*adapted from* [Theodorakopoulos (2011)]).

Before proceeding with further examples of complex melting profiles, a comment is in order here. Genomic heterogeneity destroys the sharp first order transition characteristic of the homogeneous PBD model (and of actual melting profiles of long polynucleotide chains). Remnants of multistep melting can be seen in the inset of Fig. 10.2, which suggests melting of 200-bp chunks. This is in general accord with theoretical expectations on the effect of disorder on phase transitions (reduction of effective dimensionality, hence enhancement of the role of fluctuations, rounding of an incipient transition). On the other hand, in an interesting analysis of the effects of heterogeneity within the PBD model context, it has been argued [Cule and Hwa (1997)] that multistep behavior should be self-averaging, albeit with a large crossover length varying exponentially with the ratio α/b. This would suggest a crossover length of the order of 10^9. Natural sequences however seem to contain some large scale inhomogeneities which cause traces of multistep melting to persist. An example is provided below in Fig. 10.5. Even the melting profile of the human genome [Blake *et al.* (1999)], with a size of the order of 10^9 still displays clear signs of a non-uniform sequence distribution.

10.3.2   *The pBR322 plasmid*

The plasmid pBR322, a sequence with 4361 base pairs and 53.8% GC content exhibits a rich, multipeak melting profile; this property makes it an ideal testing ground for a detailed validation of a DNA melting model. Fig. 10.3 shows the experimental profile [Delcourt and Blake (1991)] at $0.075M$ Na$^+$ ion concentration, as well as a best-fit with the heterogeneous PBD model. The fit has been obtained with Morse depth parameter values $D_{AT} = 0.1255\,eV$, $D_{GC} = 0.1655\,eV$. I note again that the two depths are the only free parameters in the fit.

The calculated melting curve, although not in perfect agreement with the experimental one, reflects the latter's complexities to a considerable extent. Note in particular the matched positions of the four main peaks. The superimposed melting map (melting temperatures, in steps of 0.5 K, uniquely defined for each base pair, cf. previous subsection) again exhibits the distinct vertical regions known from statistical (Poland–Scheraga-type) theories of DNA denaturation which coincide with peaks of the melting profile and characterize cooperative melting of large domains, extending over hundreds of base pairs. In fact, the various spatially separated domains are so finely "tuned" to melt at very slightly different temperatures

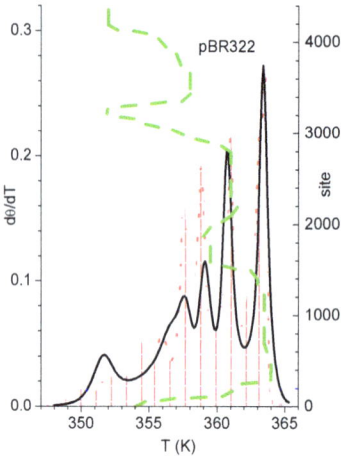

Fig. 10.3 The differential melting curve of the pBR322 plasmid. The full line shows theoretical results based on the PBD model, the dotted curve with the shaded area experimental results redrawn from [Delcourt and Blake (1991)]. Superimposed is the the melting map (dashed curve, site index on the right y-scale). Note how the vertical segments of the melting map match the peak positions of the differential melting curve. (*adapted from* [Theodorakopoulos (2010)]).

that the melting of this particular sequence exhibits a remarkable degree of reversibility[Perelroyzen *et al.* (1981)].

10.4   Predictive power of the PBD model

10.4.1   *Dependence of the Morse depths on salt concentration*

A careful analysis of the critical behavior of pure AT and GC sequences reveals essentially linear dependence of the melting temperature on the depth D of the Morse potential in the range of interest. On the other hand, it has long been known than the DNA melting temperature depends logarithmically on the concentration, c, of Na^+ ions (cf. [Blake and Delcourt (1998)] for details).

In developing a correct parametrization of the PBD model, it is therefore natural to make the *Ansatz*

$$D_\sigma = D_\sigma^0 + \kappa_\sigma \log\left(\frac{c}{c_0}\right),$$ (10.18)

where $\sigma =$ AT, GC and c_0 is some reference salt concentration. I will choose $c_0 = 0.075M$ and $D^0_{AT} = 0.1255 \ eV$, $D^0_{GC} = 0.1655 \ eV$, the parameters used to fit the pBR322 plasmid profile (section 10.3.2). The Morse depths obtained by fitting the T7 phage profile (section 10.3.1) can then be used to extract estimates of the proportionality constants, $\kappa_{AT} = 0.00855$ eV and $\kappa_{GC} = 0.00615$ eV, respectively. This allows the determination of D_{AT} and D_{GC} at any salt concentration, and therefore, in principle, the prediction of melting behavior for any long DNA sequence.

10.4.2 *Parameter-free computation of melting profiles*

In the previous subsection I showed how it is, in principle, possible to determine appropriate Morse potential depth parameters for AT and GC pairs at any Na^+ ionic strength. This enables us to compute the melting profile of any genomic sequence without making use of any freely adjustable parameters. Fig. 10.4 shows how these "predicted" melting profiles compare

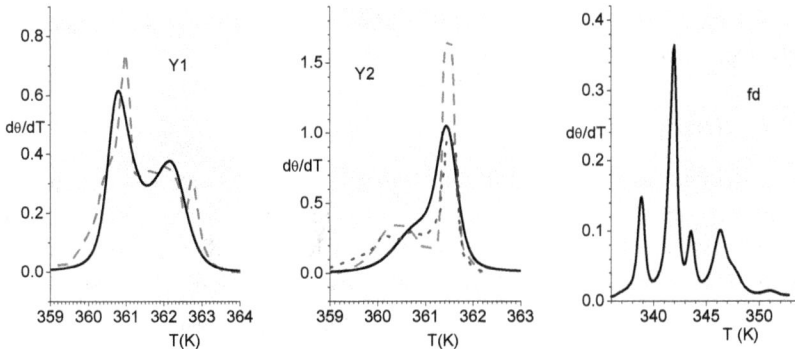

Fig. 10.4 *Left panel:* Comparison of the experimental differential melting curve of the Y1 fragment of the ϕX174 phage (2746 base pairs; dashed line, data by [Tachibana *et al.* (1982)], redrawn from [Wartell and Benight (1985)]) with theoretical profile computed from the PBD model with no adjustable parameters (full line). *Middle panel:* same for the Y2 fragment (1695 base pairs) of the ϕX174 phage; the dotted line represents another set of data by [Perelroyzen *et al.* (1981)], also redrawn from [Wartell and Benight (1985)]. *Right panel:* Experimental melting profile of the fd-phage (6408 base pairs, dashed line with shaded area, data redrawn from [Wada *et al.* (1976)]), theoretical profile (full line) computed from the PBD model with no adjustable parameters. (*adapted from* [Theodorakopoulos (2010)]).

with experiments in three different instances. The left and middle panel compare theory and experiment in the case of two distinct fragments Y1 (2746 base pairs), Y2 (1695 base pairs) of the ϕX174 bacteriophage. The Na^+ ion concentration is $c = 0.195M$ and the corresponding Morse depths, based on Eq. 10.18 are, $D_{AT} = 0.12905\,eV$ and $D_{GC} = 0.16805\,eV$, respectively. The right panel illustrates the case of the fd-bacteriophage, which consists of 6408 base pairs and features a multipeak structure. In this case the Na^+ ion concentration is $c = 0.0195M$, as in the case of the T7- phage (cf. above), therefore the same Morse parameters can be used.

In all three cases considered, PBD-based calculations predict melting temperatures and profiles of long, genomic sequences in considerable detail.

10.4.3 *Longer sequences*

The numerical method described in this chapter can deal with much longer sequences. As an example, Fig. 10.5 displays the numerically computed melting profile of the bacterial endosymbiont *Carsonella ruddii*. The sequence consists of 159662 base pairs with a GC content of 16.6 % [Nakabachi *et al.* (2006)] and the profile has been computed using the Morse depth parameters for a Na^+ concentration 0.195M. Note that even at this genomic length clear traces of multistep melting persist.

10.4.4 *Unzipping*

It is possible, using the methods of this chapter, to describe unzipping of genomic samples, as performed in typical experiments which exhibit sensitivity to local sequence composition. Here, I will confine myself to showing that the PBD model, with the parametrization of section 10.4.1, captures the essential features of the unzipping process, as described in 8.2.6 for homogeneous chains. There, it was shown that the unzipping force f_c is related to the characteristic decay length, $\xi_\perp$, of the bound state eigenfunction of the TI equation via

$$f_c = \frac{k_B T}{\xi_\perp}. \tag{10.19}$$

The absolute value of the numerically computed relevant eigenfunction at 300 K is plotted, on a logarithmic scale, in Fig. 10.6. The parameters used correspond to pure AT and pure GC composition. The unzipping forces are, respectively, 5 pN and 8.8 pN, i.e. the required force for GC unzipping is about twice as large as that for AT unzipping. The ratio is in fair

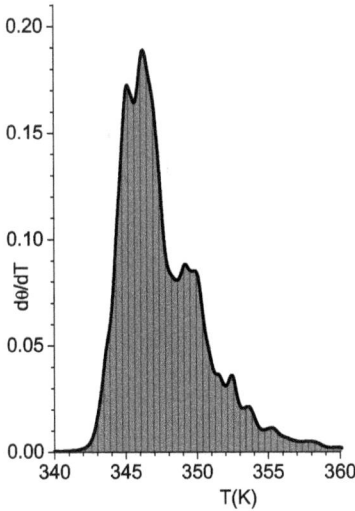

Fig. 10.5 Numerically computed melting profile of the bacterial endosymbiont *Carsonella ruddii*, 159 662 bps, GC content 16.6%, at salt concentration 0.195M (*adapted from* [Theodorakopoulos (2010)]).

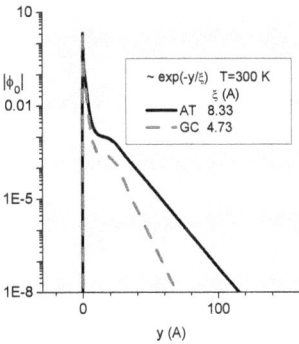

Fig. 10.6 The absolute value of the bound eigenfunction for the PBD model at room temperatures for a pure AT and a pure GC chain, numerically computed with Morse depth parameters corresponding to a salt concentration 0.075M. The limiting exponential form provides estimates of the transverse correlation length $\xi_\perp$, and, consequently, of the unzipping force $f = k_B T/\xi_\perp$ (8.2.6).

agreement with experiment [Rief *et al.* (1999)], although the magnitudes of the experimentally determined unzipping forces, 9 and 20 pN, respectively, were larger by about a factor of 2.

The characteristic decay length diverges as

$$\xi_\perp \propto (T_c - T)^{-1/2}$$

in the case of the PBD model [Theodorakopoulos *et al.* (2000)]. Consequently, the model captures correctly the force-temperature phase diagram of Fig. 6.2, where the phase boundary approaches the temperature axis with an infinite slope.

Chapter 11

Temperature dependent DNA flexibility

11.1 Introduction

Early light scattering studies[Peterlin (1953a,b)] performed prior to the discovery of the double helix already revealed that DNA in solution was a stiff molecule at a microscopic scale, with an estimated persistence length of the order of 300 Å. Subsequent measurements established base-stacking interactions as a major factor in the rigidity of duplex DNA[Hagerman (1988)]. Low salt concentration is insofar a factor as it tends to favor further rigidity; at Na^+ ion concentrations higher than $.01M$, the persistence length seems to saturate at values around 500 Å[Hagerman (1988)]. Single molecule experiments (cf. Section 2.5) confirmed that the elastic behavior of long chains at room temperature follow the predictions of wormlike chain model (WLC).

A systematic study of the temperature dependence of flexibility of long genomic samples based on sedimentation experiments [Gray and Hearst (1968)] concluded that, up to about 50 °C, the persistence length decreases with increasing temperature roughly as $1/T$, in accordance with (1.13). Alternatively, one may also interpret the result of [Gray and Hearst (1968)] as stating that the stiffness parameter κ remains approximately constant over the temperature range examined.

There are at least two caveats to this overall picture of broadly constant, WLC-obeying, bending stiffness of DNA duplexes. One of them regards relatively short sequences (< 200 base pairs), where the presence of even one or two imperfections (e.g. "nicks"arising from permanent bends) may introduce measurable changes in the effectively observed persistence length. I will address this issue separately in the next chapter. The second caveat has to do with the strongly fluctuating premelting and melting regimes where bubble-like local openings abound and are expected to decrease bending stiffness. In order to discuss them we would need an extension of the Kratky–Porod Hamiltonian (1.3),

$$H_{KP} = -\sum_{j=1}^{N-1} J_{j,j+1}\vec{t}_j \cdot \vec{t}_{j+1} \qquad (11.1)$$

with site-dependent stiffness parameters to account for inhomogeneities — e.g. possible soft spots. It should be of course born in mind that such an approach — and this is in fact a third caveat — neglects the direct effects of torsional elasticity. This however will be the price of conceptual simplicity.

11.2 Statistical mechanics of the heterogeneous KP chain

11.2.1 *The partition function*

The partition function of the heterogeneous KP chain,

$$Z_N = \int d\Omega_1 \cdots d\Omega_N \prod_{j=1}^{N-1} e^{b_j(\vec{t}_j \cdot \vec{t}_{j+1} - 1)}$$

$$= 4\pi \prod_{j=1}^{N-1} [4\pi i_0(b_j)] \qquad (11.2)$$

where $b_j = \beta J_{j,j+1}$, turns out to be exactly factorizable. This can be done because the integrations over the solid angles can be successively performed, exactly as in the homogeneous case (cf. Eqs. 1.5, 1.6), each integration giving rise to a factor $4\pi i_0$, which now however depends on the local stiffness constant.

11.2.2 *Correlations*

Correlation functions also turn out to be factorizable. Repeated application of (1.10) is again possible and leads to

$$< \vec{t}_i \cdot \vec{t}_{i+n} > = \prod_{j=1}^{n} \mu_{i+j}, \qquad (11.3)$$

where $\mu_j = \coth b_j - 1/b_j$. Owing to the lack of translational invariance however, an effective persistence length can now only be defined as an average of the decay of correlations over the initial site. Equivalently, one may calculate the average of a global quantity, e.g. the square of the end-to-end distance and invert the relationship (1.18) to define an effective persistence length for the heterogeneous chain. I will adopt the second approach.

11.2.3 The second moment

The second moment of the end-to-end distribution function satisfies (1.15), as in the homogeneous case, i..e.

$$< R_N^2 > /a^2 = \sum_{j=1, i=1}^{N} < \vec{t}_i \cdot \vec{t}_j >$$

$$= N + 2 \sum_{j=2}^{N} \sum_{i=1}^{j-1} < \vec{t}_i \cdot \vec{t}_j >, \qquad (11.4)$$

where I have separated out the $i = j$ terms in the double summation.

Each one of the terms in the double summation is given explicitly by a product of the type (11.3). In principle therefore one can compute the resulting second moment numerically. There is however a problem with such a "brute force" computation because it involves $\mathcal{O}(N^3)$ operations, a number which can easily become astronomical for genomic length sequences. Fortunately, there is a computationally efficient solution, which I illustrate because it will be of use elsewhere.

Let

$$\Phi_1 = 0,$$

$$\Phi_j = \sum_{i=1}^{j-1} < \vec{t}_i \cdot \vec{t}_j >, \quad j = 2, 3, \cdots, N. \qquad (11.5)$$

The various Φ_j's obey the following recursive relationships:

$$\Phi_2 = < \vec{t}_1 \cdot \vec{t}_2 > = (\Phi_1 + 1)\mu_2,$$

$$\Phi_3 = < \vec{t}_1 \cdot \vec{t}_3 > + < \vec{t}_2 \cdot \vec{t}_3 > = \mu_2 \mu_3 + \mu_3 = (\Phi_2 + 1)\mu_3,$$

and, in general,

$$\Phi_j = (\Phi_{j-1} + 1)\mu_j, \quad j = 2, 3, \cdots, N. \qquad (11.6)$$

Computation of all the Φ_j's involves $\mathcal{O}(N)$ recursive steps, after which a single summation gives

$$< R_N^2 > /a^2 = N + 2 \sum_{j=2}^{N} \Phi_j. \qquad (11.7)$$

In other words, the computation of the second moment has been reduced to an $\mathcal{O}(N)$ numerical algorithm.

11.3 Flexibility and melting

11.3.1 *Magnetic birefringence of DNA in solution*

Each base pair in a duplex defines a local plane perpendicular to the double helical axis. Base pairs are known to be magnetically anisotropic. In other words, the magnetic susceptibilities $\chi_{||}$, parallel to the helical axis, and $\chi_\perp$, perpendicular to the helical axis, differ in magnitude. The difference $\Delta\chi = \chi_{||} - \chi_\perp$ is negative, i.e. each pair behaves diamagnetically. The extra energy

$$-\frac{1}{2}\Delta\chi \sum_{j=1}^{N}(\vec{B}\cdot\vec{t}_j)^2 \tag{11.8}$$

should be added to (11.1) and results in a slight reorientation of the basal planes (less than 1% at 12 Tesla) perpendicularly to the applied magnetic field $\vec{B}$.

This slight reorientation of base planes was exploited in an elegant experiment [Maret *et al.* (1975)] which measured the magnetically induced *optical birefringence* $\Delta n = n_{||} - n_\perp$ of the refractive index along directions parallel and perpendicular to the field. Optical birefringence occurs because base pairs are anisotropic in their *electrical polarizability* properties as well, leading to an anisotropic dielectric response of the medium at optical frequencies. Standard electromagnetic theory implies

$$\Delta n = \frac{2\pi}{n}\,\rho\,\Delta\alpha\,\hat{Q} \tag{11.9}$$

where n is the average index of refraction, $\Delta\alpha = \alpha_{||} - \alpha_\perp$ the anisotropy of the electronic polarizability tensor of a single base pair in the molecular frame, ρ the number of monomers (base pairs) per unit volume, and

$$\hat{Q} = \frac{1}{N}\sum_{j=1}^{N}\frac{1}{2}\langle\{3\cos^2\theta_j - 1\}\rangle, \tag{11.10}$$

where $\cos\theta_j \equiv \vec{t}_j\cdot\vec{B}/|\vec{B}|$, is an averaged orientational factor whose origin can be traced to the anisotropy of the axially symmetric polarizability tensor in the laboratory frame. The average in (11.10) must be performed in the full magnetic Kratky–Porod Hamiltonian, i.e. including the term (11.8); in the absence of a magnetic field it would vanish. On the other hand, "full magnetic" in this instance means "leading order". The reason is that even at relatively high magnetic fields, the magnetic energy $\Delta\chi B^2$

is much smaller than the typical thermal energies $k_B T$. Accordingly,

$$\hat{Q} = \Delta\chi \frac{B^2}{2k_B T} Q, \qquad \text{where}$$

$$Q = \frac{1}{N} \sum_{i,j=1}^{N} \langle \frac{1}{2}(3\cos^2\theta_j - 1)\cos^2\theta_i \rangle_0 \qquad (11.11)$$

and the 0 subscript implies averaging over field-free configurations of the KP chain (11.1). The measured birefringence is therefore directly proportional to the orientational factor (11.11):

$$\Delta n = \frac{2\pi}{n} \rho \, \Delta\alpha \, \Delta\chi \frac{B^2}{2k_B T} Q. \qquad (11.12)$$

For a homogeneous KP chain of ν monomers, where $J_{j,j+1} = J \, \forall \, j$, the double sum in (11.11) is [Wilson (1978); Hagmann *et al.* (2009)]

$$\frac{2}{15} \left\{ \nu \frac{1+u}{1-u} - 2\frac{u(1-u^\nu)}{(1-u)^2} \right\} \qquad (11.13)$$

where $u = 1 - (3/K)(\coth K - 1/K)$, $K = J/k_B T$.

At room temperatures, where the persistence length is much larger than the monomer distance $a = 3.4\mathring{A}$, I recognize (cf. (1.13)) that $K \approx \lambda/a \gg 1$. To leading order in $1/K$, $u \approx 1 - 3/K$ and the first term in the brackets of (11.13) dominates over the second, resulting in

$$Q \approx \frac{4}{45} \cdot \frac{\lambda}{a}. \qquad (11.14)$$

Looking at Fig. 11.1, we see that the measured value of the birefringence at 20 C corresponds to $Q \approx 15$[1]. This leads to a value of $\lambda \approx 57nm$, roughly in line with other estimates of the DNA duplex persistence length. Fig. 11.1 also confirms that for temperatures lower than 40–50 °C, the dependence of the orientational factor Q — and therefore of the persistence length — on temperature remains weak, consistent with the $1/T$ power law.

11.3.2 *Birefringence in the premelting and melting regime*

As the temperature increases beyond 50 C, base pair openings begin to proliferate. A first estimate of opening probabilities can be obtained by the

---

[1]In order to extract Q from the birefringence data of ([Maret *et al.* (1975)]) it is necessary to have estimates of the anisotropy of the diamagnetic susceptibility and the electronic polarizability (per base pair), as well as the mean index of refraction n. [Theodorakopoulos and Peyrard (2012)] use $n = 1.33$ [Maret and Weill (1983)], $\Delta\chi = -1.55 \times 10^{-20}$ erg/T^2 [Bryce *et al.* (2004)] and $\Delta\alpha = -18.2\mathring{A}^3$ [Stellwagen (1981)]

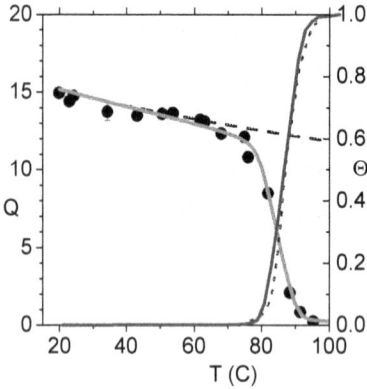

Fig. 11.1 Magnetic DNA birefringence in solution. The orientational average Q (11.11) as a function of temperature (left y-scale). Filled circles: experimental data from a sample of calf thymus DNA [Maret *et al.* (1975)] converted to Q values; solid curve, calculated according to (11.15); dashed curve, homogeneous ds-DNA. Also shown is the melting fraction (right-y scale) according to PBD theory (solid line) and experiment [Maret *et al.* (1975)] (dashed line). (*adapted from* [Theodorakopoulos and Peyrard (2012)]).

local melting fraction. We do not expect a single base pair opening to cause local softening. If, however, two consecutive base pairs open at the same time, it is reasonable to expect a softening of the corresponding stiffness constant. On this basis it is possible to formulate an Ansatz that can be used with the heterogeneous KP Hamiltonian (11.1). We assume that the local stiffness constant will take values between the double-stranded, hard limit $J = 7.02 \times 10^{-12} erg$, and the single-stranded, soft limit $J' = J/50$, according to

$$J_{j,j+1} = (1 - P_0^{j,j+1})J + P_0^{j,j+1} J', \tag{11.15}$$

where $P_0^{j,j+1}$ is the joint probability that base pairs (j) and $(j + 1)$ are open. This joint probability is a sequence-dependent quantity which can be readily calculated by applying the techniques of Chapter 10. Using the sequence of a 290 kbp long segment of a bovine chromosome [2] with a similar melting temperature and setting the Na^+ concentration equal to $0.165M$ in (10.18), as in the experiment of [Maret *et al.* (1975)], and no adjustable parameters, results in excellent agreement with experiment (cf. Fig. 11.1) in regard to both the orientational factor Q and the melting fraction.

---

[2] Gene Bank, Bos taurus breed Hereford chromosome 12 (NW_001848864.1)

Some of the properties of the sequence used in the theoretical analysis [Theodorakopoulos and Peyrard (2012)] are summarized in Fig. 11.2. The average size of an intact cluster is estimated to be of the order of many thousands of base pairs at room temperatures. The quantity P_0/Θ expresses the conditional probability that if any given base pair is in the open state, the next one will also be in the open state. At room temperatures P_0/Θ is of the order of 0.1, much larger than the a priory probability Θ of being in the open state. In other words there is a *bubble aggregation* tendency.

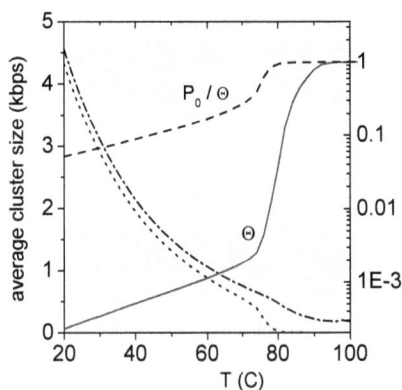

Fig. 11.2 Some average properties of the 290 kbp long sequence from Bos taurus breed Hereford chromosome 12 (*adapted from* [Theodorakopoulos and Peyrard (2012)]). Left yscale: the average size (in bps) of intact clusters (dash-dotted curve); the inverse of the melting fraction (dotted curve). Right y scale: the melting fraction Θ (solid curve) and the conditional probability P_0/Θ that if a base pair is in the open state, the next one will also be in the open state (dashed curve); note that the latter quantity is much larger than Θ (bubble aggregation tendency).

The statistical tendency of bubbles to aggregate is ultimately responsible for the nonlinear relationship between melting fraction and softening. Fig. 11.1 shows that Q, or the *effective persistence length* drops by e.g. 25% when the melting fraction is only .1. Visible softening occurs already at the premelting regime.

It is worth noting that there is at least one other long known macroscopic manifestation of how a relatively small number of fluctuational openings can change DNA behavior significantly prior to the occurrence of proper melting. The *viscosity* has been observed [Freund and Bernardi (1963)] to decrease dramatically in the premelting regime.

11.4   Enhanced flexibility of a short sequence at elevated temperatures

A similar picture emerges from another set of temperature-dependent measurements of the cyclization rates of λ-phage DNA fragments. Recall that
the rate of cyclization, or ring closure, can be calculated exactly in the
framework of the WLC model (cf. 1.4.0.2). In fact it is possible to generalize the WLC model to include details of the helical structure [Shimada
and Yamakawa (1984)]; this results in a pattern of small oscillations reflecting the pitch of the double helix superimposed on the curve of Fig.
1.5. Measurements of the cyclization rate of 200 bps long fragments could
therefore be translated into very precise estimates of the persistence length
[Geggier *et al.* (2011)]. The results are displayed in Fig. 11.3 and show
a hitherto unexpectedly strong temperature dependence setting off above
room temperature.

It is in fact relatively straightforward to show that the bubble-based
scenario developed in the previous section to account for the birefringence
data is consistent with the cyclization measurements as well. In order to
demonstrate this I consider the second moment of the end-to-end distance
distribution function. This is a quantity that can be calculated numerically for any heterogeneous KP chain, following the computational rules
of 11.2.3. Here, this implies using the bubble-related Ansatz (11.15) for
the stiffness constants; the joint probabilities $P_0^{j,j+1}$ are calculated from
the PBD model with the actual sequence information and the experimental Na^+ molar concentration $c = 0.004$ in (10.18)[Theodorakopoulos and
Peyrard (2012)]. The persistence length can then be estimated from $< R^2 >$
using the relationship (1.18)

$$< R^2 >= 2\lambda L - 2\lambda^2(1 - e^{-L/\lambda}),$$

where $L = 200 \times 0.34 = 68$ nm is the contour length of the fragment. The
result, calculated for $J' = 0.14 \times 10^{-12}$ erg and $J = 6.13 \times 10^{-12}$ erg, is
shown in Fig. 11.3 and appears consistent with the experimental findings.

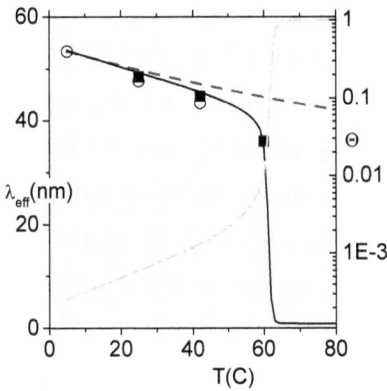

Fig. 11.3 Temperature dependence of DNA persistence length (*reproduced from* [Theodorakopoulos and Peyrard (2012)]). Experimental data (filled squares, open circles) obtained [Geggier *et al.* (2011)] by measuring the ring closure probability of a 200 bps long λ-phage fragment. The solid curve shows the estimate obtained using a heterogeneous KP chain with stiff and soft joints distributed according to (11.15) and bubble probabilities from a PBD calculation. Also shown is a constant stiffness, $1/T$, reference curve (dashed line), and the melting fraction (dashed-dotted line, right y-axis) obtained from the PBD calculation. Note how at 40 C a melting fraction of the order of 10^{-3} corresponds to an almost 10% drop in measured stiffness.

Chapter 12

Is DNA softer at the 100-nm scale?

According to the picture presented so far, DNA at the scale of a few hundred nm should appear at room temperature essentially as a stiff rod. However, experiments in the last ten years have cast considerable doubts on the general validity of such a sweeping statement. Cyclization experiments with some short sequences ($\leq$ 200 bps)[Cloutier and Widom (2004, 2005); Vafabakhsh and Ha (2012)] have measured a significantly higher rate of ring closure than that expected from WLC model calculations with a 50 nm persistence length. SAXS-based measurements of even shorter sequences, 10-35bp [Mathew-Fenn et al. (2008)] and 15-89 bp [Yuan et al. (2008)] also suggest a notably softer structure at this scale.

In order to analyze elastic behavior at this short scale it will be necessary to consider explicitly the role of inhomogeneities. A hint of the potential significance of this was given in the last chapter where the relevance of bubbles in temperature-dependent elasticity is concerned. A further hint is implicit in the strongly nonlinear dependence of average stiffness on bubble population in the case of the short sequence of 11.4. In the following I will develop the theory of scattering from heterogeneous Kratky–Porod chains and then go on to describe how defects of two different types can change the detected apparent stiffness of short DNA chains.

12.1 Scattering from a heterogeneous Kratky–Porod chain

The structure factor which describes scattering from a heterogeneous chain of N segments is defined in terms of the positions of the individual monomers, just like in the homogeneous case 4.1, as

$$S(\vec{q}) = \frac{1}{N+1} + \frac{2}{(N+1)^2} \sum_{j=2}^{N} \sum_{i=0}^{j-1} S_{i,j}(\vec{q}),$$

where the individual terms
$$S_{i,j}(\vec{q}) = <e^{i\vec{q}\cdot(\vec{R}_i - \vec{R}_j)}>$$
$$= \int d\vec{r}\, e^{i\vec{q}\cdot\vec{r}} P_N(\vec{r}|\vec{R}_i - \vec{R}_j)$$
are Fourier transforms of the probability distribution of the distance between the ith and the jth mononer, $P_N(\vec{r}|\vec{R}_i - \vec{R}_j)$.

12.1.1 *The probability distribution for the distance between any two monomers*

The probability distribution $P_N(\vec{r}|\vec{R}_i - \vec{R}_j)$, $\quad i < j$ is a generalization of the end-to-end distance probability distribution. Just like (1.28), it can be expressed as a ratio of two multiple integrals. The point to notice is that contributions from segments $> j$ or $< i$ are the same in both numerator and denominator (cf. the identical argument for the homogeneous KP chain in 4.2.1). Therefore
$$P_N(\vec{r}|\vec{R}_i - \vec{R}_j) = P_{j-i}(\vec{r}|\vec{R}_i - \vec{R}_j), \tag{12.1}$$
i.e. the distribution function - or its Fourier transform - must only be computed for the chain with $j - i$ segments connecting the ith to the jth mononer. In other words we need the end-to-end distribution for the heterogeneous chain, a generalization of (1.36),
$$P_{j-i}(\vec{q}) = \left(\prod_{k=i}^{j-1} \mathbf{F}^{(k)}\right)_{00} \tag{12.2}$$
where the matrix $\mathbf{F}$ has been defined in (1.34) and the superscript k implies, in the notation of 11.2.1, that it is a function of the dimensionless stiffness constants $b_k = \beta J_{k,k+1}$. Consequently, the probability distribution function $P_N(\vec{r}|\vec{R}_i - \vec{R}_j)$ will now depend on all stiffness constants b_k for k between i and j.

12.1.1.1 *The end-to-end distance probability distribution for the heterogeneous KP chain*

An immediate corollary of the above result is an exact expression for the Fourier transform of the end-to-end distance probability distribution for the heterogeneous KP chain,
$$P_N(\vec{q}) = \left(\prod_{k=i}^{N} \mathbf{F}^{(k)}\right)_{00}, \tag{12.3}$$
which is computable by direct matrix multiplication and can be Fourier-transformed to provide the real-space distribution $P_N(r)$.

12.1.2 *The structure factor of the heterogeneous KP chain*

It is now possible to generalize the steps outlined in 11.2.3 using matrix, rather than scalar iterates, and evaluate the double sum 12.1. Let

$$f_j = \sum_{i=0}^{j-1} S_{i,j},$$

suppressing for a moment the dependence on $\vec{q}$ and the interaction strengths. The required sum in (12.1 is a single sum

$$\sum_{j=2}^{N} f_j.$$

Now it is possible to rewrite the terms $f_2 = S_{1,2}$, $f_3 = S_{1,3} + S_{2,3}$, $f_4 = S_{1,4} + S_{2,4} + S_{3,4}$ as

$$f_2 = \mathbf{F}_{00}^{(1)},$$

$$f_3 = \left(\mathbf{F}^{(1)}\mathbf{F}^{(2)} + \mathbf{F}^{(2)}\right)_{00} = \left(\left[\mathbf{F}^{(1)} + \mathbf{I}\right]\mathbf{F}^{(2)}\right)_{00},$$

$$f_4 = \left(\mathbf{F}^{(1)}\mathbf{F}^{(2)}\mathbf{F}^{(3)} + \mathbf{F}^{(2)}\mathbf{F}^{(3)} + \mathbf{F}^{(3)}\right)_{00} = \left(\left[\mathbf{F}^{(1)}\mathbf{F}^{(2)} + \mathbf{F}^{(2)} + \mathbf{I}\right]\mathbf{F}^{(3)}\right)_{00}.$$

Using the matrices defined by the iteration

$$\mathbf{\Phi}^{(j+1)} = \left(\mathbf{\Phi}^{(j)} + \mathbf{I}\right)\mathbf{F}^{(j)}, \quad j = 1, 2, \cdots, N - 1, \tag{12.4}$$

with $\mathbf{\Phi}^{(1)} = \mathbf{0}$ a matrix whose elements all vanish, one can now to express the structure factor of the heterogeneous KP chain in the form

$$S(\vec{q}) = \frac{1}{N+1} + \frac{2}{(N+1)^2} \sum_{j=2}^{N} \mathbf{\Phi}_{00}^{(j)}. \tag{12.5}$$

In summary, the steps outlined above provide an $\mathcal{O}(N)$ computational procedure for the structure factor of a heterogeneous Kratky–Porod chain.

The quality of the computational procedure is compared to a corresponding MC calculation, with 30000 conformations, in Fig. 12.1. The system chosen consists of 100 monomers. All coupling constants except the middle 3 have the value $b = 145$. The middle 3 have the value $b' = 3$. This corresponds roughly to a normally "hard" DNA, with a nominal persistence length of 49 nm, interrupted by a 3-bp bubble, characterized by a coupling constant appropriate to ss-DNA. The matrices $\mathbf{F}$ and $\mathbf{\Phi}$ are truncated at dimensionality 21. Results from the two types of computation are essentially identical.

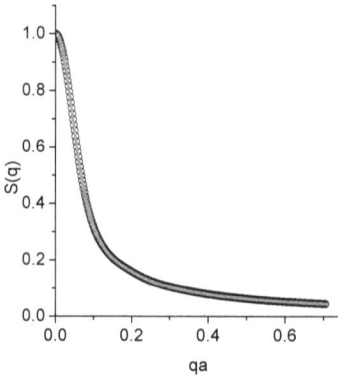

Fig. 12.1   The structure factor of an inhomogeneous KP chain, computed from (12.5) (continuous line) and from a MC calculation with 30000 conformations (open circles). The chain consists of 100 monomers. Coupling constants are equal to $b_j = 145$ for all j except $b_{49} = b_{50} = b_{51} = 3$. The matrix $\mathbf{F}$ is truncated at dimensionality 21.

12.2   Hinge-induced apparent softening of a 100-bp sequence

I can now look at how the presence of a single soft link, a "hinge" connecting the two stiff halves at midpoint, affects the observed elastic properties of short DNA chains. The physical origin of the soft link may be e.g. thermal activation of a bubble. Consider the example introduced in the previous section, a 100 bp sequence with a 3-bp pair bubble at the middle. The dimensionless stiffness constants are

$$b_j = \begin{cases} 3 & \text{if } j = 49, 50, 51; \\ 145 & \text{otherwise.} \end{cases}$$

The left panel of Fig. 12.2 shows the structure factor for the chain with the 3-link bubble, as well as the "reference" chain without any soft links. Note that, although a difference between the two structure factors exists, it is not very pronounced, and therefore presents a priori a challenge to the experiment seeking to detect it directly. A somewhat better visual contrast is provided by the Kratky plot of the quantity $qS(q)$ displayed in the right panel of Fig. 12.2 but, again, the excess intensity in the vicinity of the bump is of the order of 10%.

The Guinier analysis of the long-wavelength regime (inset, left panel of Fig. 12.2), provides estimates of the radius of gyration. Again, the difference between homogeneous ($R_g/a = 23.5$) and hinge-at-midpoint

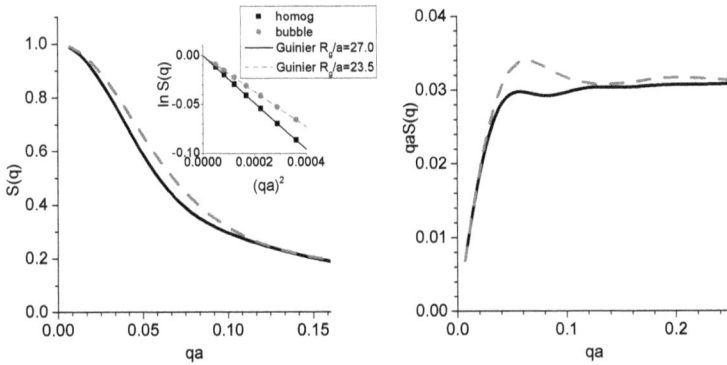

Fig. 12.2 Scattering "signature" of a hinge. *Left panel:* $S(q)$ for a chain with 99 segments (continuous line) and (i) all stiffness constants $b_j = 145 \; \forall j$, (ii) with a hinge (3-link bubble) in the middle, $b_j = 145 \; \forall j \neq 49, 50, 51$, $b_j = 3 \; \text{for} j = 49, 50, 51$ (dashed line). The inset displays a Guinier analysis of the long-wavelength regime. The extracted radius of gyration is, in units of the monomer distance, 27.0 in the homogeneously stiff case and 23.5 in the case where a bubble is present. *Right panel:* a Kratky plot of the product $qS(q)$ in the two cases described in the left panel. The signature of the bubble is a "hump" around $qR_g \sim 1$.

$(R_g/a = 27.0)$ cases is of the order of 15%. It is then possible to use the discrete version of the Benoit–Doty relationship (1.24) in order to determine the persistence length. In the case of the inhomogeneous chain with the bubble in the middle, the persistence length should be understood as an effective one. The procedure is displayed in Fig. 12.3. Owing to the form of the R_g vs λ curve, a relatively small variation in R_g results in a dramatic change in λ. In this particular case, the estimated λ/a changes from 142.9 to 41.8. In physical units this translates (with $a = 0.34$ nm) to a change from 49 to 14 nm. A flexible hinge formed by e.g. a 3-site bubble is therefore, in principle, a mechanism which could in principle account for drastic softening of a short DNA molecule. Nonetheless, as the analysis of chapter 11 suggests, bubble excitation probabilities are rather small at room temperature and would therefore be unlikely to bring about the complete flexibility required in the hinge model calculation of the present section. A similar conclusion has been reached by a systematic, bubble-based analysis of cyclization data [Forties *et al.* (2009)].

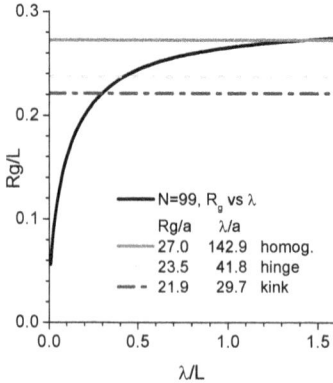

Fig. 12.3   Dependence of the radius of gyration on the persistence length for an (effectively homogeneous) KP chain with $N = 99$ segments, using (1.24) and $\lambda/a = \coth b - 1/b$. A relatively small variation in the observed radius of gyration results in a dramatic change in the estimate of the persistence length. The solid line is drawn at the R_g found in the homogeneous case. The dotted and the dashed-dotted lines denote, respectively, the R_g's found in the hinge and kink cases. Note how the hinge and the kink result in a reduction of the effective persistence length by a factor of 3.4 and 4.8 respectively.

12.3   Permanent bends and effective softening of short sequences

An alternative to the flexible hinge considered in the previous section is a permanent local bend of the chain, also known in the literature as a nick or a kink. Kinks were suggested long ago [Crick and Klug (1975)] in connection with chromatin folding, as possible metastable structures that could occur without a drastic distortion of the 3-dimensional DNA backbone. An appropriate generalization of the KP model to describe such a bend would be to set the total conformational energy equal to

$$H = -J \sum_{j=1}^{N-1} \cos(\theta_j - \theta_j^0) \tag{12.6}$$

where $\cos\theta_j = \vec{t}_j \cdot \vec{t}_{j+1}$. Setting all θ_j^0's equal to zero would reduce (12.6) to the standard KP model. A nonzero θ_j^0 describes the preference of the jth link to form an angle θ_0 rather than orient the jth and $(j+1)$th segments parallel to each other.

The thermodynamics of (12.6) cannot be calculated exactly. However, its properties at thermal equilibrium can be readily obtained by a MC calculation.

I now consider the case of a $\pi/2$ permanent bend, or kink, at the middle of a KP chain with 100 monomers. The change in the structure factor, compared with the homogeneous case, as obtained from a MC simulation, is displayed in Fig. 12.4. The left panel shows that there is a well defined excess intensity, better visible in the Kratky plot in the right panel. The Guinier analysis shows that the radius of gyration decreases from 27.0 to 21.9 monomer distances, which, using the procedure of the previous section translates to a persistence length of 29.7 monomer distances (Fig. 12.3). The presence of the kink makes the KP chain appear effectively 4.8 times softer!

Another way to view the drastic effect of the $\pi/2$ kink on the elastic properties of the short KP chain is by looking at the end-to-end distance probability distribution. Fig. 12.5 shows both cases, with and without the kink. In summary therefore, it is quite clear that permanent bends present an interesting alternative scenario for effective softening of a KP chain. In the next section I will discuss how this scenario plays in the case of real DNA in solution.

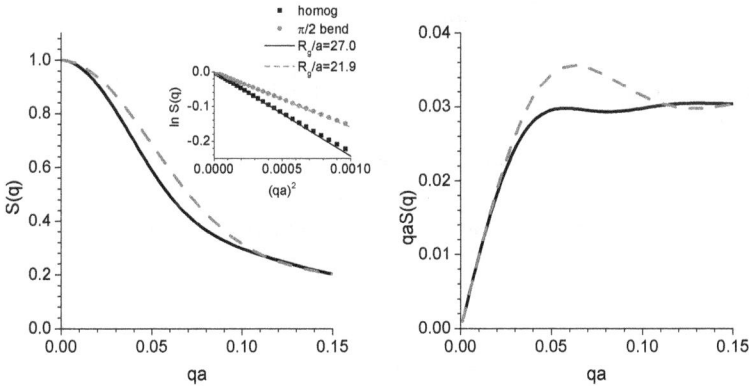

Fig. 12.4 Scattering "signature" of a $\pi/2$ kink (permanent bend) as obtained by a MC calculation (30000 conformations) of the model (12.6). *Left panel:* $S(q)$ for a homogeneous KP chain with 99 segments (continuous line) and (i) all stiffness constants $b_j = 145\ \forall j$, (ii) with the same stiffness constants and a permanent bend in the middle, $\theta_{49}^0 = \pi/2$ and zero for all other j's (dashed line). The inset displays a Guinier analysis of the long-wavelength regime. The extracted radius of gyration is, in units of the monomer distance, 27.0 in the homogeneously stiff case and 19.2 in the case where a kink is present. *Right panel:* a Kratky plot of the product $qS(q)$ in the two cases described in the left panel. The signature of the kink is a pronounced "hump" around $qR_g \sim 1$.

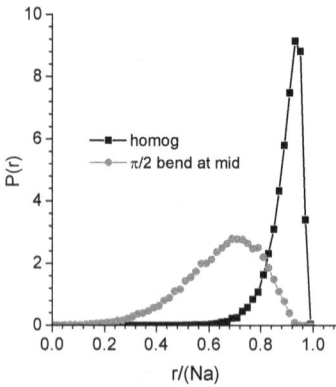

Fig. 12.5   Probability distribution of the end-to-end distance for a chain with 100 monomers, obtained by a MC calculation (30000 conformations). The squares denote the result for a stiff KP chain, the circles for the same chain with a $\pi/2$ kink in the middle (parameters as in the previous figure).

12.4   Kinky DNA in solution

In this section I will outline some results from a small-angle scattering experiment [Schindler *et al.* (2018)] performed with a 145 bp long DNA molecule containing the so called "601" strong positioning (Widom) sequence known for easy wrapping around a histone core in nucleosome formation. The results complement previous evidence of apparent DNA softening at the 100 bp scale and the data analysis included a detailed statistical model search for bend-like defects. The search, performed in the context of a more realistic model of DNA elasticity (including torsional modes), locates sites with a high probability of bending and thus provides support for the original Crick–Klug kink hypothesis. In the following, I will give a very simplified version of the analysis in terms of the extended KP model (12.6), which nonetheless provides a strong plausibility argument for the presence of a permanent kink-like bend.

Fig. 12.6 (left panel) shows the measured absolute SAXS intensity as a function of wavevector. A Guinier plot, (cf. (4.5)), shown in the inset, provides the estimates $I_0 = 0.247$ cm^{-1} and $R_g = 93.3$Å for the forward scattering intensity and the radius of gyration, respectively. From the radius of gyration one can estimate the persistence length of an "equivalent KP" chain, using the procedure shown graphically in Fig. 12.3, to be $\lambda = 85$Å. The right panel shows the scattering data, reduced by the factor I_0, and compares them with two possibilities. First, the uniformly soft DNA, i.e. a homogeneous KP model with $\lambda = 85\text{Å}$ (dashed line), as obtained by the Guinier plot. Although this type of fit matches well both the high- and the low-q behavior, it clearly deviates from the data in the intermediate region. The other possibility is a stiff-KP-plus kink model (12.6) computed by the MC method; the stiffness constants are uniform, corresponding to the standard DNA persistence length, 51 nm, and the permanent bend is at link 67, $\theta_{67}^0 = 100°$. The inset shows a Kratky plot of the data and both theoretical variants. It is clear that the kink provides a better — although of course not unique — fit to the experimental data.

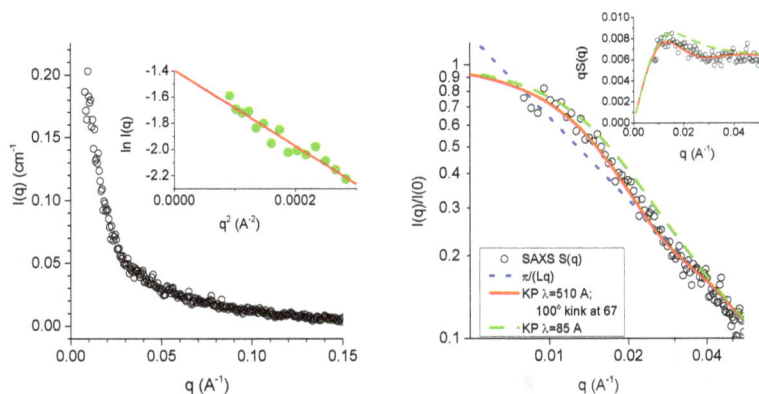

Fig. 12.6 Summary of SAXS results from the 601 Widom sequence, consisting of 145 bps. *Left panel:* absolute intensities $I(q)$ vs scattering wavevector [Schindler *et al.* (2018)]. The inset is a Guinier plot of $\ln I$ vs q^2, in the low-q regime $0.0095\text{Å}^{-1} < q < 0.017\text{Å}^{-1}$ from which estimates of $I(0) = 0.247$ cm^{-1} and $R_g = 93.3\text{Å}$ can be obtained. *Right panel:* A double logarithmic plot of the structure factor $S(q)$ vs. q. The circles represent experimental data, the solid curve is the result of a MC calculation of model (12.6) with stiffness constants $b_j = \beta J = 150.0 \,\forall j$, corresponding to $\lambda = 51$ nm, and a permanent $100°$ bend at position 67. Also shown (dotted curve) is the limit $\pi/(qL)$ and a uniformly soft ($\lambda = 85\text{Å}$) KP structure factor (dashed line). The inset is a Kratky plot of the product $qS(q)$ showing the experimental data and the same MC KP-plus-kink, and soft KP chain results.

12.5   Concluding remarks

Although the experiment described in the previous section provides compelling structural evidence for a reduced DNA rigidity at the 100-bp scale, and, at the very least, a plausible interpretation in terms of permanent (metastable) bends, the overall issue remains open and controversial. At least two systematic studies [Mastroianni *et al.* (2009); Geggier and Vologodskii (2010)] have provided estimates of bending stiffness for various sequences which are roughly consistent with the consensus value of the persistence length. In addition, early SANS measurements (cf. 4.8) also seem to be consistent with the consensus value and certainly exclude the type of behavior reported in [Schindler *et al.* (2018)]. The resolution of this puzzle will probably require a more detailed investigation of particular sequence effects. In other words, is softness at this length scale associated with very specific nucleosome folding features? And if so, what are the factors which control the (meta)stability of the (semi)permanent bends?

Chapter 13

Thermodynamic stability of DNA hairpins

13.1 Self-complementary sequences

Single strands of DNA with complementary sequences at their terminal regions will, provided that the general conditions for B-DNA hybridization are met, self assemble spontaneously in solution to form a harpin-like conformation with a double-helical stem and a single-stranded loop (Fig. 13.1). Hairpin loop conformations are ubiquitous in both ss-DNA and RNA, participating in a number of biological functions, e.g. regulation of gene expression and DNA recombination.

The thermodynamic stability of DNA hairpins is controlled by the competition between the duplex free energy of the stem and the entropic cost of maintaining the attached loop. In general, the longer the loop, the higher its entropic cost; hairpins with longer loops will generally tend to be more unstable, i.e. they unravel ("melt") at lower temperatures than those with shorter loops.

13.2 Biomolecular beacons

The hairpin structure of Fig. 13.1 has been exploited, combined with fluorescence resonance energy transfer (FRET) spectroscopy, in the development of biomolecular beacons [Tyagi and Kramer (1996)], i.e. devices which can recognize single DNA strands with a specific sequence. The principle behind the device is as follows:

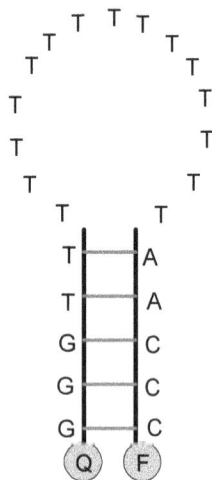

Fig. 13.1 Schematic representation of a DNA hairpin consisting of a stem with 5 base pairs and a loop with 15 bases. In single-molecule experiments utilizing fluorescence resonance energy transfer (FRET), a fluorophore (F) and a quencher (Q) are chemically attached to the opposing ends of the stem.

13.2.1 *Fluorescence resonance energy transfer (FRET) spectroscopy*

Fluorescence resonance energy transfer (FRET) is the process of nonradiative energy transfer from an excited molecular chromophore (fluorophore, donor) to another chromophore (quencher, acceptor) by means of intermolecular dipolar interaction [Förster (1946)]. Because the interaction falls off as the sixth power of the intermolecular distance, energy transfer can be triggered as a sensitive response to the proximity of donor and acceptor, inhibiting the former's fluorescence. The characteristic, Förster distance varies from 1 to 10 nm, depending on the chromophore pair. FRET spectroscopy has been widely applied to the exploration of spatial patterns of complex biological systems [Clegg (1995)] by attaching appropriate dye molecules to specific biomolecular sites.

13.2.2 *Fluorescence upon hybridization*

Fig. 13.1 displays the dye molecule arrangement in the case of a DNA hairpin. A fluorophore (F) and a quencher (Q) are chemically attached to the terminal bases of the single DNA strand. Furthermore, in general, the loop may consist of an arbitrarily complex sequence. As long as the base pairs in the stem remain bound, the quencher's proximity to the fluorophore inhibits fluorescence. The DNA hairpin with its attached dye molecules is an example of a *biomolecular beacon*[Tyagi and Kramer (1996)].

Now suppose that a number of single strands ("targets") which contain a sequence exactly complementary to the hairpin's loop is added to the solution. Hybridization will certainly occur because, at room temperature - neglecting for a moment the difference between GC and AT enthalpies - there will be a great enthalpic advantage in forming 15 base pairs by pairing to the loop segment while surrendering the 5 pairings of the stem. As this occurs, the two ends of the hairpin become spatially detached and the quencher is no more in the vicinity of the fluorophore. Consequently, fluorescence occurs, and its intensity can be calibrated to yield the degree of hybridization with the target sequence. Hybridization is efficient and highly specific: if any mismatches or single nucleotide deletions are present, it does not occur[Tyagi and Kramer (1996)].

13.3 Thermodynamic stability of hairpins

Hairpin dissociation can be detected by macroscopic means (e.g. UV absorption) which respond to the average number of bound base pairs present in the solution. On the other hand, single-molecule fluorescence resonance energy transfer (FRET) spectroscopy offers a superb tool that can follow the beacons' own thermodynamics and kinetics. DNA beacons are in this context no more just a tool for sequence recognition; they are themselves an object of study. What one expects in particular is that, by looking at the systematics of how opening and/or closing kinetics and thermodynamic stability depend on the length of the loop, a clearer picture of ss-DNA entropic elasticity might emerge.

13.3.1 *Open-closed equilibria*

The above approach was used by Libchaber and coworkers [Bonnet *et al.* (1998); Goddard *et al.* (2000)] to study the melting profiles and kinetics of hairpins with varying length. Hairpin self-assembly and decomposition can be viewed as a chemical equilibrium similar to the helix-coil transformation (5.1.2.1), with an equilibrium constant

$$K = \frac{[\text{open}]}{[\text{closed}]} \equiv e^{-\Delta G/k_B T} = e^{\Delta S/k_B - \Delta H/k_B T}$$

characterizing the ratio of molecules in the open and closed hairpin states, and an open fraction given by

$$\theta = \frac{1}{1 + e^{\Delta G/k_B T}}.$$

In the present chapter I will restrict myself to the "canonical" case of poly(T) loops, as depicted in Fig. 13.1. Poly(A) loops, although interesting in themselves, are known to have the property of "stacking"; this complicates both the dynamics and the equilibrium properties and must be considered separately.

Fig.13.2 shows melting curves obtained [Goddard *et al.* (2000)] for hairpins with poly(T) loops of three different lengths and the 5-base-pair stem of Fig. 13.1 in an aqueous solution with $0.25 M$ molar concentration of Na^+ ions. All four curves can be described as equilibria (cf. above) between the open and closed states. Moreover, the van't Hoff enthalpy is in all cases practically the same, approximately 28 Kcal/mol. For comparison, the enthalpy of the stem obtained by the nearest-neighbor Ansatz (5.5) and Table 5.1 is

$$\Delta H_s = \Delta H_{GC}^{init} + 2\Delta H_{GG/CC} + \Delta H_{GT/CA} + \Delta H_{TT/AA} + \Delta H_{AT}^{init}$$
$$= 29.9 \, \text{Kcal/mol.} \tag{13.1}$$

Since van't Hoff and thermodynamic enthalpies are essentially identical, one may safely assume that the opening/closing of the hairpin is of the AON type (cf. section 7.1.3). An elementary model of the thermodynamic transition is presented in the next subsection.

13.3.2   *Hairpin statistical mechanics: stem vs loop*

Modeling the process as being of the AON type, I assign a statistical weight of 1 to the open state. The closed state will be characterized by the weight

$$Z_{closed} = Z_{stem} Z_{loop}. \tag{13.2}$$

The statistical weight of the stem can be approximated by

$$Z_{stem} = e^{\Delta H_s/k_B T - \Delta S_s/k_B} \tag{13.3}$$

where the enthalpy ΔH_s is given by (13.1) and the entropy, by analogous application of the nearest-neighbor Ansatz (5.5) and Table 5.1, can be obtained to be $\Delta S_s = 85.7$ cal/mol/K[1].

---

[1] Note that, according to our sign convention, the enthalpy of the stem is in fact $-\Delta H_s$, and its entropy $-\Delta S_s$

The statistical weight of the loop is in essence entropic. One way to deal with it is in terms of the worm-like chain. An alternative approximation is (cf. the previous discussion of unzipping in 6.1.2) to model the loop as a FJC chain with an effective number of segments $\mathcal{N}^*$, equal to the total length divided by the Kuhn length, i.e.

$$\mathcal{N}^* = (N_L + 1) \cdot \frac{a}{2\lambda}, \tag{13.4}$$

where N_L is the number of bases in the loop, a the ss-DNA monomer distance, and λ the persistence length.

In this case, the statistical weight of the loop would be of the type

$$Z_{loop} = \frac{A}{\mathcal{N}^{*\,c}} \tag{13.5}$$

where $c = 3/2$ and A is a numerical constant. Note that, since I have arbitrarily assigned a unit weight to the open state, Z_{loop} is in fact the probability of loop formation.

In fact, the Ansatz (13.5) is more general. It can describe situations of polymer ring formation under more general conditions of excluded volume *and* local rigidity. Local rigidity has implicitly been taken into account by introducing an effective number of rigid segments, each equal to the Kuhn length. The effect of excluded volume (cf. 1.5) is typically addressed by the model class of self-avoided walks (SAW, cf. also the discussion of the entropic contributions by denatured loops in the context of the Poland–Scheraga model of DNA melting, section 7.2.3). The exponent c characterizing the ring closure, or cyclization probability, has been calculated to be 1.92 for a close-packed three-dimensional lattice model (a face-centered cubic lattice, [Martin *et al.* (1967)]). For the moment, however, I will refrain from specifying the value of the exponent c.

Combining Eqs. 13.2-13.5 I obtain the open fraction of hairpins, i.e. the quantity measured by the normalized fluorescence, as

$$\theta = \frac{1}{1 + \frac{A}{\mathcal{N}^{*\,c}} \cdot e^{\Delta H_s/k_B T - \Delta S_s/k_B}}. \tag{13.6}$$

The melting point, defined by $\theta(T_m) = 1/2$ should therefore depend on the number of bases in the loop as

$$\frac{1}{T_m} = const. + \frac{k_B}{\Delta H_s} \cdot c \, \ln(N_L + 1). \tag{13.7}$$

I have plotted the inverse melting temperatures for hairpins of varying loop lengths in Fig. 13.2, verifying the linear behavior predicted by (13.7). From the slope of the straight line and the known stem enthalpy (13.1) one obtains the exponent $c = 0.17K^{-1} \times \Delta H_s/(1000\ k_B) = 2.58 \pm 0.15$. From the intercept one can obtain the value of the numerical constant A (cf. above). Using $\lambda = 1.3$ nm (cf. 6.1.2) and $a = 0.65$ nm [Sim *et al.* (2012)] leads to $A = 1.98$. The theoretical melting profiles obtained with these values are plotted in Fig. 13.2 and are in very good agreement with the experimental data[2].

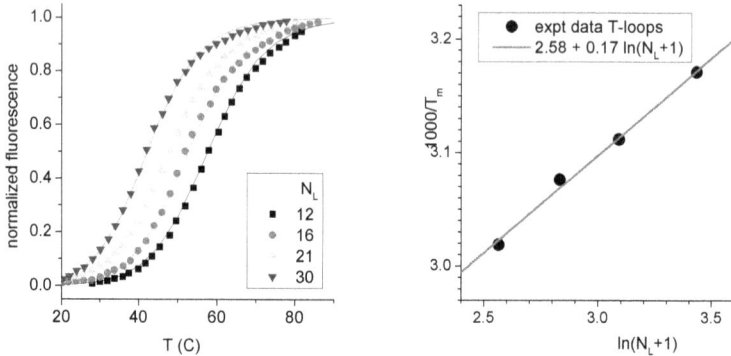

Fig. 13.2 *Left panel:* Normalized fluorescence intensity vs. temperature for hairpins of varying loop length. Symbols are data (redrawn from [Goddard *et al.* (2000)]) obtained with hairpins with stem as in Fig. 13.1, squares, circles, triangles and inverted triangles representing T-loops of length 12, 16, 21 and 30, respectively. Continuous lines are theoretical computations, based on (13.6) with an exponent $c = 2.58$. *Right panel:* Dependence of the inverse melting temperature on the number of bases in the loop.

13.4 Excluded volume, SAWs and electrostatic repulsion in ss-DNA

The experimentally determined value of the cyclization probability exponent $c = 2.58$ exceeds significantly the value 1.92 predicted by the asymptotics of SAW models. Can this discrepancy be understood?

Before proceeding any further, it will be useful to recapitulate what the FJC model with the excluded volume constraint describes. With the parameter choice made above $2\lambda/a = 4$, a single FJC segment describes a total of 4 segments joining the original bases of ss-DNA. This means that

---

[2]The quality of the fit depends very little on the exact choice of the persistence length, which simply changes the value of the numerical constant A. The decisive parameter is the exponent c, determined essentially by the slope of the straight line of Fig. 13.2.

in the FJC description the stem now effectively consists of a single rigid segment at the start and another one at the end of the chain. The number of segments in the loop is given by (13.4). Cyclization is therefore defined not by the proximity of monomers at the start and end of the chain, but as an *internal* approach of the second to the next-to-last monomer.

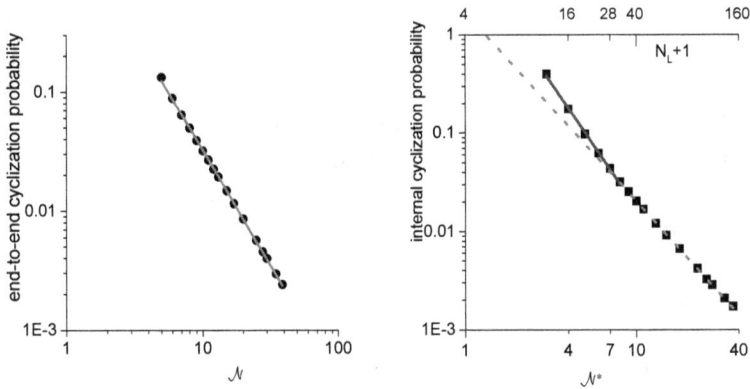

Fig. 13.3 *Left panel:* Cyclization probability of the FJC with excluded volume vs the number of segments $\mathcal{N}$. Symbols are MC data, obtained with an excluded volume parameter $D/a = 1$(cf. 1.5). The continuous line represents a least-square fit with a slope -1.926 ± 0.010. *Right panel:* Internal cyclization probability of the same chain forming a loop with $\mathcal{N}^* = \mathcal{N} - 2$ segments and a stem with a single pair of segments. The upper axis is calibrated to show the number $N_L + 1$, where N_L is the number of bases in the actual hairpin loop. This calibration depends on the value of the parameter $2\lambda/a = 4$. The fit to the lowest six points, $\mathcal{N}^* \leq 8$, yields a a straight line with slope -2.57 ± 0.05 (solid line). For $\mathcal{N}^* \geq 9$ the slope (dotted line) is -1.92, same as for end-to-end cyclization.

The statistics of internal ring closure has been studied in the framework of the SAW problem. A renormalization group calculation [des Cloizeaux, J. (1980)] gives $c = 2.18$ for an fcc lattice. Whereas this serves as a guide, it should be borne in mind that the SAW results describe the asymptotic properties of very large systems. The DNA hairpins under consideration are by definition very small systems, for which asymptotic behavior may not — indeed is not expected to — have settled.

It is therefore useful to perform a direct MC calculation for the small chains to determine the probability of internal ring closure. Because of the small size, it is possible, in spite of the high rejection rate, to perform a "brute force" MC calculation. The results are summarized in Fig. 13.3. The left panel shows that the end-to-end cyclization proceeds exactly according to the asymptotic SAW scenario, i.e. with an exponent $c = 1.926$. The right panel describes the probability of internal ring closure, i.e. a proximity of the second to the next-to-last monomer. A straight line fit to the points with $\mathcal{N}^* \leq 8$, corresponding to $N_L \leq 31$, i.e. exactly in the range of the experimental data, yields $c = 2.57 \pm 0.05$, in complete agreement with the findings of the previous section. The points corresponding to longer chains are described by $c = 1.92$; in other words, for longer chains asymptotic behavior would prevail; in that limit, internal ring closure would become indistinguishable from end-to-end ring closure. Although the actual extent of the agreement between theory and experimental data on hairpin thermodynamic stability may be somewhat fortuitous, the relevance of excluded volume effects to the conformations of short chains of single-stranded DNA has already been systematically demonstrated. A comprehensive study using small-angle X-ray scattering (SAXS) techniques under varying salt concentrations [Sim *et al.* (2012)] showed a systematic variation of the effective polymer size (as measured by the radius of gyration) with Na^+ concentration. The scaling exponent ν (cf. 1.5.2) was found to have non-canonical values (higher than 0.6) at low salt concentrations, dropping to a value slightly lower than 0.6 at salt concentrations of 1M, as the ions in the solvent effectively screen out the long-range electrostatic repulsion which normally swells ss-DNA. The effect was observed in both poly(T) and poly(A) chains, but ν remains significantly larger than 0.6 in the case of poly(A), even at the highest salt concentrations. In contrast, poly-T at the highest salt concentration is an almost "canonical" polymer with excluded-volume interaction.

Appendix A

Monte Carlo simulations of the Kratky–Porod chain

It is relatively straightforward to generate random configurations of the KP chain with N segments 1. The chain can be "grown" as follows: The first unit vector $\vec{t}_1$ defines the z-direction. The second vector is generated by randomly specifying a polar and an azimuthal angle, θ_1 and ϕ_1, respectively, relative to the first vector. The process is repeated $N-1$ times, i.e. until the Nth segment vector has been specified. The set of angles $\{\theta_j, \phi_j\}, j = 1, \cdots, N-1$ describe a unique configuration of the N-segment chain.

In order to perform averages in the canonical ensemble defined by the KP model 1.4. a particular configuration must be accepted or declined according to its probability of occurrence. Since we are building configurations in a step-by-step function, this criterion must also be applied at each vectorial step. This means in effect that (i) owing to the rotational symmetry of the model, any azimuthal angle in the interval $0 \leq \phi_j < 2\pi$ is equally acceptable and (ii) a Metropolis[Metropolis et al. (1953)] (Monte Carlo, MC) rule should applied to decide on the acceptability of any given polar angle. According to the KP model, the energy cost of any nonzero polar angle is

$$\Delta\epsilon = \frac{\kappa}{a}(1 - \cos\theta_j).$$

The Metropolis rule states that, at any given temperature T,

- if

$$\frac{\Delta\epsilon}{k_B T} = \frac{\lambda}{a}(1 - \cos\theta_j) \leq 1,$$

the random choice in the interval $0 \leq \theta_j < \pi$ is accepted[1], and the jth vector of the particular configuration is generated;

---

[1]In fact, in order to conform with the requirements of phase space measure, cf. 1.2.1, the choice of random numbers should be such that the resulting $\cos\theta_j$ are uniformly distributed in the interval $(-1, 1)$.

- if on the other hand

$$\frac{\Delta \epsilon}{k_B T} = \frac{\lambda}{a}(1 - \cos \theta_j) > 1,$$

the choice is rejected and another random number θ_j must be generated.

The configurations thus generated can be used to study equilibrium averages of any of their features in the canonical ensemble. For example, one may look at the end-to-end distance, compute it for a large numbers of configurations, and plot the resulting histogram. The results of the numerical simulation can then be compared to the theoretical predictions.

So far, I have described how a MC simulation for a discrete, KP model works. Let me now add a practical note on how the parameters of a simulation - which is always performed on a finite, discrete chain - must be chosen in order to yield information relevant to the WLC limit. Note first that a WLC is completely characterized by the ratio λ/L of its persistence length to its contour length. A WLC simulation must choose a discrete KP chain satisfying two requirements. First, the discretization scale must be sufficiently fine to approximate a continuum. This means not only that the chain must have a large number of segments N, but also that the distance between successive monomers must be a very small fraction of the persistence length, typically smaller than $1/10$. Second, for a given size N, the stiffness constant must be chosen in such a way that it reproduces the ratio of persistence length to contour length specified by the WLC limit. This requires in effect a dimensionless stiffness $b \equiv (\kappa/a)/(k_B T) \approx \lambda/a = (\lambda/L)N$. A more precise correspondence is provided by the numerical solution of (1.12):

$$\coth b - \frac{1}{b} = e^{-a/\lambda} = e^{-(L/\lambda)/N}. \tag{A.1}$$

For stiff chains, where λ and L are of comparable size, this criterion implies a high microscopic stiffness, which in turn will lead to a high rejection rate of the MC rule. Simulations will then take longer and longer. Fig.A.1 provides examples of MC simulations for the end-to-end distance distribution function of the WLC performed at varying discretization scales.

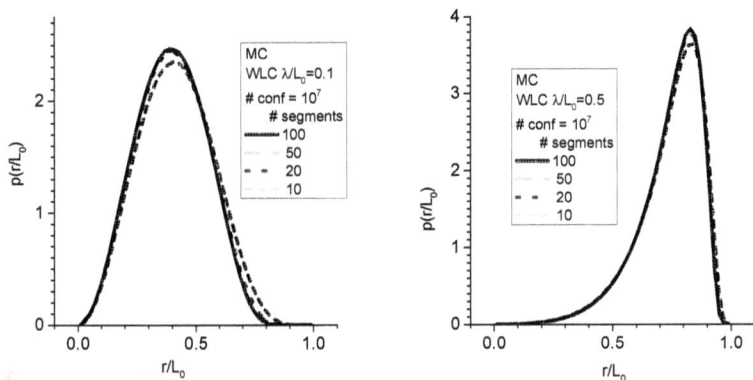

Fig. A.1 *Left panel:* MC simulation of the end-to-end distance distribution of a "flexible" WLC chain, $\lambda/L_0 = 0.1$. Results are shown for 4 different simulations, at varying discretization scales, with $N = 100, 50, 20, 10$ segments and a KP stiffness chosen according to (A.1). Each simulation consists of 10^7 configurations of the corresponding KP chain. *Right panel:* same for a "stiffer" WLC chain, $\lambda/L_0 = 0.5$

Appendix B

Landau's theorem on the absence of phase transitions in one-dimensional systems

In words, Landau's theorem [Landau and Lifshitz (1980)] states that *macroscopic phase coexistence cannot occur at finite temperatures in one dimensional systems*. The proof is outlined below.

Consider a one-dimensional system with N sites, which may exist in either phase A or phase B. Let θ be the fraction of phase A; furthermore, let there be $m << N$ contacts between the phases, each of energy ϵ. Depending on the details of the mathematical model, these contacts can be steplike (Ising) or continuous domain walls characterizing the transition regions between domains of two different phase types. The sites of the m contacts can be selected from the total number of sites in

$$\frac{N!}{m!(N-m)!}$$

distinct ways. The alternation of phase A and phase B domains thus generates a (dimensionless) entropy associated with contacts (*domain wall entropy*)

$$S_{DW}(m, N) = \ln\left[\frac{N!}{m!(N-m)!}\right] \approx m \ln\left[\frac{Ne}{m}\right], \tag{B.1}$$

where I have used Stirling's approximation for the factorial function.

The free energy of a particular configuration of the total system with m transition regions can be thought of as consisting of three terms. One describes regions of phase type A, another describes regions of phase type B, and a third term has its origin in the transition regions (domain walls)

$$F(m, N) = N\theta f_A + N(1-\theta)f_B + F_{DW}(m) \tag{B.2}$$

where

$$F_{DW}(m, N) = m\epsilon - k_B T S_{DW}(m, N). \tag{B.3}$$

At thermodynamic equilibrium, the number of transition regions is determined by the minimization condition

$$\frac{\partial}{\partial m} F(m, N) = 0 \qquad (B.4)$$

which results in

$$m = \bar{m} = N e^{-\epsilon/(k_B T)} \, . \qquad (B.5)$$

According to (B.5) the number of transition regions is macroscopic (of order N), i.e. the density of transition regions is finite. The system breaks up into $\bar{m}$ regions (domains) of finite size $e^{\epsilon/(k_B T)}$. Macroscopic phase separation can only occur at zero temperature (as the domain size goes to infinity). Landau's argument covers a wide range of one-dimensional systems, e.g. double-well on-site potentials (Ising universality class), or periodic on-site potentials. Notable exceptions include certain systems with long-range interactions, as well as the Peyrard–Bishop model of DNA denaturation, presented in Section 8.2. In all such cases the domain wall energy ϵ becomes unbounded as system size goes to infinity [Theodorakopoulos (2006)] and Landau's theorem becomes inapplicable.

Appendix C

Dynamical theory of DNA melting: The soliton analogy

C.1 Soliton-like field configurations of the Peyrard–Bishop model

The PB Hamiltonian (8.1) generates the dynamical equations of motion

$$\mu \ddot{y}_n = k[y_{n+1} + y_{n-1} - 2y_n] - \frac{\partial V}{\partial y_n}, \qquad (C.1)$$

or, in the continuum limit used in section 8.2.3.5,

$$\ddot{y} = c_0^2 \frac{\partial^2 y}{\partial x^2} - \frac{1}{\mu} \frac{\partial V}{\partial y}, \qquad (C.2)$$

where $c_0 = (k/\mu)^{1/2} a = \omega_0 a$.

Eq. C.2 admits a uniform, static solution

$$y_0(x) = 0 \quad \forall x. \qquad (C.3)$$

This corresponds to all sites being in their absolute equilibrium positions at the bottom of the Morse potential well. There is, however, another pair of exact, static solutions of (C.2) [Dauxois *et al.* (2002)],

$$y^{\pm}(x) = \frac{1}{\alpha} \ln[1 + e^{\pm(x-x_0)/d}] \qquad (C.4)$$

where $d/a = (k/D)^{1/2}/\alpha$. Both solutions represent exact nonlinear field configurations which "interpolate" between the stable and metastable equilibrium states of the Morse potential, at $y = 0$ and $y = \infty$, respectively. For example, the $(+)$ solution vanishes if $x \ll x_0$ and grows proportionately with x if $x \gg 0$. Both the elastic and the on-site contributions to the energy vanish in the first case and equal D per site in the second case. Accordingly, if lattice sites are numbered from 0 to N, we can write

$$E^+(x_0) = 2D \left(N - \frac{x_0}{a} \right) + \mathcal{O}(N^0) \qquad (C.5)$$

for the energy of the (+) solution centered at x_0. In other words, the total energy grows linearly with the number of sites whose displacements lie on the flat part of the Morse potential, or, using the DNA language, where base pair separation is large (molten state). Nonlinear static field configurations which interpolate between distinct "vacuum" states are well known in field theory as kinks, domain walls, or topological solitons. If the energies of the vacua are the same, as is the case e.g. if $V(y) = (1/2)(y^2 - 1)^2$, with the equivalent minima at $y = \pm 1$, the energies of the nonlinear configurations (kinks) arise from a local distortion in the interpolation region and are finite. This is not the case with the Morse potential, where the energy (C.5) of the soliton-like solution (C.4) can be of the order of the system size.[2]

A three-dimensional representation of the field (C.4) including the on-site potentials is shown in Fig. C.1.

In the absence of thermal fluctuations, the field configuration (C.4) is not stable. Owing to the property

$$E^+(x_0 + a) - E^+(x_0) = -2D, \tag{C.6}$$

the configuration will spontaneously move to the right by one lattice site; this will be successively repeated until the unbound portion of the chain has been entirely "zipped" back.

C.2   Thermal fluctuations and soliton stability

In order to examine the relative stability of the two competing equilibrium states (C.3) and (C.4) at finite temperature one must examine their free energy difference. The difference arises because of different entropic contributions from sites to the right of x_0 (looking at the y_0^+ static solution, vs. the uniform static solution $y_s = 0$). In the case of the uniform solution, all sites are characterized by small oscillations at optical phonon frequencies (cf.(8.7))

$$\Omega_q^{opt} = [4\omega_0^2 \sin^2(qa/2) + \Omega_0^2]^{1/2}. \tag{C.7}$$

In the case of the y_0^+ small oscillations behave differently at the right and left sides of x_0. At the left of x_0, the oscillations take place around the absolute minimum $y = 0$, i.e. they are again optical phonons; at the right

---

[2]It is of course technically correct to argue that in the case of the PB Hamiltonian the interpolation region is, strictly, of infinite extent, since the displacement field, large as it may be, never becomes really infinite.

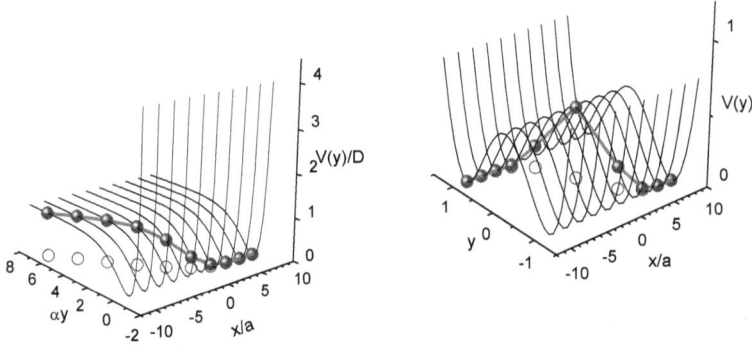

Fig. C.1   *Left panel:* Three-dimensional representation of the soliton-like field configuration in the case of the PB model. The projection on the (y, n) plane (open circles)
represents the solution y^- (C.4). The vertical axis represents the local on-site potential
energy of the Morse potential. Each unbound site contributes an amount D to the on-
site energy. An equal amount is contributed from the elastic field energy. *Right panel:*
For comparison, the kink-like soliton of the ϕ^4 potential, $V(y) = (1/2)(y^2 - 1)^2$. The
energy minima are equivalent. Contributions to the soliton energy originate only in the
interpolation region.

of x_0 however, oscillations take place at the plateau of the Morse potential,
i.e. they are gapless, acoustic phonons (cf. Fig. 8.1) with a dispersion

$$\Omega_q^{ac} = 2\omega_0 \sin(qa/2) \tag{C.8}$$

This produces, for each site at the right of x_0, an extra amount of
entropy [Dauxois *et al.* (2002)]

$$
\begin{aligned}
\Delta s^* &= \frac{1}{N} \sum_{\nu=1}^{N} \ln \left(\frac{\Omega_{q\nu}^{opt}}{\Omega_{q\nu}^{ac}} \right) \\
&= \frac{1}{2\pi} \int_0^\pi dx \ln \left(\frac{4\omega_0^2 \sin^2(x/2) + \Omega_0^2}{4\omega_0^2 \sin^2(x/2)} \right) \\
&= \ln \left(\sqrt{1 + \frac{r}{2}} + \sqrt{\frac{r}{2}} \right),
\end{aligned}
\tag{C.9}
$$

where $r = \Omega_0^2/(2\omega_0^2) = D\alpha^2/k$ is a dimensionless measure of the relative
strengths of on-site vs. nonlocal restoring forces. The entropy (C.9) which

accompanies the generation of the domain well is positive and therefore, as the temperature is raised, it will increasingly compensate the energy cost $2D$ per site, associated with DW formation. The free energy associated with the DW

$$\Delta F_{DW}^{+} = (N - \frac{x_0}{a})(2D - k_B T \Delta s^*) \tag{C.10}$$

will vanish at

$$T_c = \frac{D}{k_B} \cdot \frac{2}{\ln\left(\sqrt{1 + \frac{r}{2}} + \sqrt{\frac{r}{2}}\right)} \tag{C.11}$$

and become negative at higher temperatures.

Note that, in the continuum limit $r \ll 1$ the above result for the critical temperature reduces to (8.38), which has also been derived in the continuum limit by using the Schrödinger equation approach. In fact, the agreement between the transfer-integral and the soliton-based approach to the thermodynamic instability has been verified for higher values of r, well into the discrete limit.[Dauxois *et al.* (2002); Theodorakopoulos *et al.* (2004)].

It should be further noted that, since the DW energy C.5 is macroscopic, the prohibition imposed by Landau's theorem on thermodynamic phase transitions in one-dimensional systems (cf. Appendix B) is not applicable in this case.

In summary, the properties of the exact, static, nonlinear solution (C.4) provide us with an alternative view of the DNA melting transition within the context of the PB model. In addition, they clarify the status of the one-dimensional phase transition of the PB model as being consistent with Landau's theorem.

Appendix D

Numerical solution of the transfer integral equation

D.1 Gauss–Legendre quadratures

The transfer integral (TI) equation can be numerically solved by transforming it to an approximately equivalent matrix eigenvalue problem. The numerical procedure is based on the general method of approximating an integral by Gaussian quadratures, using some set of orthogonal polynomials. For example [Abramowitz and Stegun (1964)], the integral of a function $f(x)$ which is non-singular in the interval $(-1, 1)$ can be approximated by a finite sum over a given set of points $\{x_i, i = 1, \cdots, n\}$,

$$\int_{-1}^{1} dx\, f(x) = \sum_{i=1}^{n} w_i f(x_i) + R_n \tag{D.1}$$

where x_i is the ith zero of the Legendre polynomial $P_n(x)$, the weights are given by the formula

$$w_i = \frac{2}{1 - x_i^2} [P_i'(x_i)]^2, \quad i = 1, \cdots, n, \tag{D.2}$$

and an error bound

$$R_n = \frac{2^{2n+1}(n!)^4}{(2n+1)[(2n)!]^3} f^{(2n)}(\xi), \quad -1 < \xi < 1 \tag{D.3}$$

(Gauss–Legendre (GL) quadratures).

Eq. (D.1) can be generalized to apply to any finite interval (a, b),

$$\int_{a}^{b} dy\, f(y) \approx \frac{b-a}{2} \sum_{i=1}^{n} w_i f(y_i) \tag{D.4}$$

where the mesh is now given by

$$y_i = \frac{b+a}{2} + \frac{b-a}{2} \cdot x_i, \quad i = 1, \cdots, n. \tag{D.5}$$

167

Both sets of numbers, $\{x_i\}, \{w_i\}, i = 1, \cdots, n$, are well tabulated in reference works and can also be readily generated by standard available numerical subroutines.

The error involved in the numerical approximation described above has been estimated to be proportional to the $(2n)$th derivative of the function f, with a numerical prefactor proportional to $n^{-1}[e/(4n)]^{2n}$ for large values of n, i.e. rapidly converging to zero. Although the GL procedure generally produces convergent results, it is not always easy to translate the theoretical error bound (D.3) to a realistic error estimate. In any numerical application it is therefore customary to ascertain that the results approach a well defined limit as $n \to \infty$ and that the estimated error tends to zero in a controlled fashion.

D.2   Application to the TI equation

It is now straightforward to apply the techniques of the previous section to approximate the TI eigenvalue problems for the PB and PBD models, (8.20 and (9.3) respectively, by a real, symmetric matrix eigenvalue problem of the form

$$\sum_{j=1}^{n} \hat{T}_{i,j} \hat{\phi}_j^{(\nu)} = \Lambda_\nu \hat{\phi}_i^{(\nu)}, \quad i = 1, \cdots, n, \tag{D.6}$$

where $\nu = 1, 2, \cdots, n$, the mesh is defined in the finite interval (y_{min}, L), and the matrix is related to the symmetrized TI kernel by

$$\hat{T}_{i,j} = \frac{L - y_{min}}{2} \cdot \sqrt{w_i w_j}\, T(y_i, y_j). \tag{D.7}$$

The eigenvalues $\{\Lambda_\nu\}$ and the eigenvectors $\{\hat{\phi}^{(\nu)}\}$ of (D.6) can be evaluated numerically by making use of readily available mathematical subroutines.

Bibliography

Abramowitz, M. and Stegun, I. (1964). *Handbook of Mathematical Functions* (Dover).

Allawi, H. T. and SantaLucia, J. (1997). Thermodynamics and NMR of internal GT mismatches in DNA, *Biochemistry* **36**, pp. 10581–10594.

Azbel, M. Y. (1979). Phase transitions in DNA, *Physical Review A* **10**, pp. 1671–1684.

Bauer, D. W., Huffman, J. B., Homa, F. L., and Evilevitch, A. (2013). Herpes virus genome, the pressure is on, *Journal of the American Chemical Society* **135**, pp. 11216–11221.

Benoit, H. and Doty, P. (1953). Light scattering from non-Gaussian chains, *Journal of Physical Chemistry* **57**, pp. 958–963.

Blake, R. D., Bizzaro, J. W., Blake, J. D., Day, G. R., Delcourt, S. G., Knowles, J., Marx, K. A., and SantaLucia, J. (1999). Statistical mechanical simulation of polymeric DNA melting with MELTSIM, *Bioinformatics* **15**, pp. 370–375.

Blake, R. D. and Delcourt, S. G. (1998). Thermal stability of DNA, *Nucleic Acids Research* **26**, pp. 3323–3332.

Bonnet, G., Krichevsky, O., and Libchaber, A. (1998). Kinetics of conformational fluctuations in DNA hairpin-loops, *Proceedings of the National Academy of Sciences* **95**, pp. 8602–8606.

Breslauer, K. J., Franks, R., Blockers, H., and Marky, L. A. (1986). Predicting DNA duplex stability from the base sequence, *Proceedings of the National Academy of Sciences* **83**, pp. 3746–3750.

Bryce, D. L., Boisbouvier, J., and Bax, A. (2004). Experimental and theoretical determination of nucleic acid magnetic susceptibility: Importance for the study of dynamics by field-induced residual dipolar couplings, *Journal of the American Chemical Society* **126**, pp. 10821–10822.

Bustamante, C., Marko, J. F., Siggia, E. D., and Smith, S. B. (1994). Entropic elasticity of lambda phage DNA, *Science* **265**, pp. 1599–1600.

Bustamante, C., Smith, S. B., Liphardt, J., and Smith, D. (2000). Single-molecule studies of DNA mechanics, *Current Opinion in Structural Biology* **10**, pp. 279–285.

Clausen-Schaumann, H., Rief, M., Tolksdorf, C., and Gaub, H. E. (2000). Mechanical stability of single DNA molecules, *Biophysical Journal* **78**, 4, pp. 1997–2007.

Clegg, R. M. (1995). Fluorescence resonance energy transfer, *Current Opinion in Biotechnology* **6**, pp. 103–110.

Cloutier, T. E. and Widom, J. (2004). Spontaneous sharp bending of double-stranded DNA, *Molecular Cell* **14**, pp. 355–362.

Cloutier, T. E. and Widom, J. (2005). DNA twisting flexibility and the formation of sharply looped protein-DNA complexes, *Proceedings of the National Academy of Sciences* **102**, pp. 3645–3650.

Cluzel, P., Lebrun, A., Heller, C., Lavery, R., Viovy, J. L., Chatenay, D., and Caron, F. (1996). DNA: an extensible molecule, *Science* **271**, pp. 792–794.

Cohen, G. and Eisenberg, H. (1966). Conformation studies on the sodium and cesium salts of calf thymus deoxyribonucleic acid (DNA), *Biopolymers* **4**, pp. 429–440.

Creighton, T. E. (1992). *Proteins* (W.H. Freeman).

Crick, F. H. C. and Klug, A. (1975). Kinky helix, *Nature* **255**, pp. 530–533.

Cule, D. and Hwa, T. (1997). Denaturation of heterogeneous DNA, *Physical Review Letters* **79**, pp. 2375–2378.

Dauxois, T., Peyrard, M., and Bishop, A. R. (1993). Entropy driven DNA denaturation, *Physical Review E* **47**, pp. R44–R47.

Dauxois, T., Theodorakopoulos, N., and Peyrard, M. (2002). Thermodynamic instabilities in one dimension: correlations, scaling and solitons, *Journal of Statistical Physics* **107**, pp. 869–891.

Dayantis, J. and Palierne, J. (1991). Monte Carlo precise determination of the end-to-end distribution function of self-avoiding walks on the simple-cubic lattice, *The Journal of Chemical Physics* **95**, pp. 6088–6099.

Delcourt, S. G. and Blake, R. D. (1991). Stacking energies in DNA, *Journal of Biological Chemistry* **266**, pp. 15160–15169.

des Cloizeaux, J. (1974). Lagrangian theory for a self-avoiding random chain, *Physical Review A* **10**, pp. 1665–1669.

des Cloizeaux, J. (1980). Short range correlation between elements of a long polymer in a good solvent, *Journal de Physique, France* **41**, pp. 223–238.

Earnshaw, W. C. and Casjens, S. R. (1980). DNA packaging by the double-stranded DNA bacteriophages, *Cell* **21**, pp. 319–331.

Errami, J., Peyrard, M., and Theodorakopoulos, N. (2007). Modeling DNA beacons at the mesoscopic scale, *The European Physical Journal E* **23**, pp. 397–411.

Essevaz-Roulet, B., Bockelmann, U., and Heslot, F. (1997). Mechanical separation of the complementary strands of DNA, *Proceedings of the National Academy of Sciences* **94**, pp. 11935–11940.

Fisher, M. E. (1964). Magnetism in one-dimensional systems — the Heisenberg model for infinite spin, *American Journal of Physics* **32**, pp. 343–346.

Fisher, M. E. (1966a). Effect of excluded volume on phase transitions in biopolymers, *Journal of Chemical Physics* **45**, pp. 1469–1473.

Fisher, M. E. (1966b). Shape of a self-avoiding walk or polymer chain, *The Journal of Chemical Physics* **44**, pp. 616–622.

Fisher, M. E. (1967). The theory of condensation and the critical point, *Physics* **3**, pp. 255–283.

Flory, P. (1953). *Principles of Polymer Chemistry* (Cornell University Press).

Förster, T. (1946). Energiewanderung und Fluoreszenz, *Naturwissenschaften* **6**, pp. 166–175.

Forties, R. A., Bundschuh, R., and Poirier, M. G. (2009). The flexibility of locally melted DNA, *Nucleic Acids Research* **37**, pp. 4580–4586.

Freund, A.-M. and Bernardi, G. (1963). Viscosity of deoxyribonucleic acid solutions in the sub-melting temperature range, *Nature* **200**, pp. 1318–1319.

Gao, Y., Devi-Prasad, K. V., and Prohofsky, E. W. (1984). A self-consistent microscopic theory of hydrogen bond melting with application to poly(dG).poly(dC), *Journal of Chemical Physics* **80**, pp. 6291–6298.

Geggier, S., Kotlyar, A., and Vologodskii, A. (2011). Temperature dependence of DNA persistence length, *Nucleic Acids Research* **39**, pp. 1419–1426.

Geggier, S. and Vologodskii, A. (2010). Sequence dependence of DNA bending rigidity, *Proceedings of the National Academy of Sciences* **107**, pp. 15421–15426.

Goddard, N. L., Bonnet, G., Krichevsky, O., and Libchaber, A. (2000). Sequence dependent rigidity of single stranded DNA, *Physical Review Letters* **85**, pp. 2400–2403.

Godfrey, J. E. and Eisenberg, H. (1976). The flexibility of low molecular weight double-stranded DNA as a function of length: II. Light scattering measurements and the estimation of persistence lengths from light scattering, sedimentation and viscosity, *Biophysical Chemistry* **5**, pp. 301–318.

Gradshteyn, I. S. and Ryzhik, I. M. (2007). *Table of Integrals, Series, and Products* (Academic Press).

Gray, H. and Hearst, J. (1968). Flexibility of native DNA from the sedimentation behavior as a function of molecular weight and temperature, *Journal of Molecular Biology* **35**, pp. 111–129.

Gueron, M., Kochoyan, M., and Leroy, J.-L. (1987). A single mode of DNA base-pair opening drives imino proton exchange, *Nature* **328**, pp. 89–92.

Hagerman, P. J. (1988). Flexibility of DNA, *Annual Review of Biophysics and Biophysical Chemistry* **17**, pp. 265–286.

Hagmann, J.-G., Kozlowski, K., Theodorakopoulos, N., and Peyrard, M. (2009). On 4-point correlation functions in simple polymer models, *Journal of Statistical Mechanics* , pp. P04011–1–19.

Harpst, J. (1980). Analysis of low angle light scattering results from T7 DNA, *Biophysical Chemistry* **11**, pp. 295–302.

Hugel, T., Rief, M., Seitz, M., Gaub, H. E., and Netz, R. R. (2005). Highly stretched single polymers: Atomic-force-microscope experiments versus ab-initio theory, *Phys. Rev. Lett.* **94**, p. 048301.

Inman, R. B. and Baldwin, R. L. (1964). Helix-random coil transitions in DNA homopolymer pairs, *Journal of Molecular Biology* **8**, pp. 452–469.

Jolly, D. J. and Campbell, A. M. (1972). Light-scattering studies on deoxyribonucleic acid flexibility: The solution properties of a small circular deoxyribonucleic acid molecule, *Biochemichal Journal* **130**, pp. 1019–1028.

Kadanoff, L. P., Götze, W., Hamblen, D., Hecht, R., Lewis, E. A. S., Palciauskas, V. V., Rayl, M., Swift, J., Aspnes, D., and Kane, J. (1967). Static phenomena near critical points: Theory and experiment, *Reviews of Modern Physics* **39**, pp. 395–431.

Kafri, Y., Mukamel, D., and Peliti, L. (2000). Why is the DNA denaturation transition first order? *Physical Review Letters* **85**, pp. 4988–4991.

Kroll, D. M. and Lipowsky, R. (1983). Universality classes for the critical wetting transition in two dimensions, *Physical Review B* **28**, pp. 5273–5280.

Landau, L. D. and Lifshitz, E. M. (1977). *Quantum Mechanics* (Pergamon).

Landau, L. D. and Lifshitz, E. M. (1980). *Statistical Physics* (Pergamon).

Latulippe, D. R. and Zydney, A. L. (2010). Radius of gyration of plasmid DNA isoforms from static light scattering, *Biotechnology and Bioengineering* **107**, pp. 134–142.

Lederer, H., May, R., Kjems, J., Baer, G., and Heumann, H. (1986). Solution structure of a short DNA fragment studied by neutron scattering, *European Journal of Biochemistry* **161**, pp. 191–196.

Leikin, S., Parsegian, V. A., and Rau, D. C. (1993). Hydration forces, *Annual Review of Physical Chemistry* **44**, pp. 369–395.

Lifson, S. (1964). Partition functions of linear chain molecules, *Journal of Chemical Physics* **40**, pp. 3705–3710.

Lipowsky, R. (1985). Critical effects at complete wetting, *Physical Review B* **32**, pp. 1731–1750.

Lubensky, D. K. and Nelson, D. R. (2000). Pulling pinned polymers and unzipping DNA, *Physical Review Letters* **85**, pp. 1572–1575.

Lubensky, D. K. and Nelson, D. R. (2002). Single molecule statistics and the polynucleotide unzipping transition, *Physical Review E* **65**, p. 031917.

Lyubchenko, Y. L., Frank-Kamenetskii, M. D., Vologodskii, A. V., Lazurkin, Y. S., and Gause, G. G. J. (1976). Fine structure of DNA melting curves, *Biopolymers* **15**, pp. 1019–1036.

Maret, G., Oldenbourg, R., Winterling, G., Dransfeld, K., and Rupprecht, A. (1979). Velocity of high-frequency sound waves in oriented DNA fibers and films determined by Brillouin scattering, *Colloid and Polymer Science* **257**, pp. 1017–1020.

Maret, G., v. Schickfus, M., Meyer, A., and Dransfeld, K. (1975). Orientation of nucleic acids in high magnetic fields, *Physical Review Letters* **35**, pp. 397–400.

Maret, G. and Weill, G. (1983). Magnetic birefringence study of the electrostatic and intrinsic persistence length of DNA, *Biopolymers* **22**, pp. 2727–2744.

Marko, J. F. (1997). Stretching must twist DNA, *Europhysics Letters* **38**, pp. 183–188.

Marko, J. F. and Cocco, S. (2003). The micromechanics of DNA, *Physics World* **16**, 3, pp. 37–41.

Marmur, J. and Doty, P. (1960). Determination of the base composition of deoxyribonucleic acid from its thermal denaturation temperature, *Journal of Molecular Biology* **5**, pp. 109–118.

Martin, J. L., Sykes, M. F., and Hioe, F. T. (1967). Probability of initial ring closure for selfavoiding walks on the facecentered cubic and triangular lattices, *The Journal of Chemical Physics* **46**, pp. 3478–3481.

Mastroianni, A., Sivak, D. A., Geissler, P. L., and Alivisatos, A. P. (2009). Probing the conformational distributions of subpersistence length DNA, *Biophysical Journal* **97**, pp. 1408–1417.

Mathew-Fenn, R. S., Das, R., and Harbury, P. A. B. (2008). Remeasuring the double helix, *Science* **322**, pp. 446–449.

Metropolis, N., Rosenbluth, A. W., Rosenbluth, M. N., Teller, A. H., and Teller, E. (1953). Equation of state calculations by fast computing machines, *Journal of Chemical Physics* **21**, pp. 1087–1092.

Mihardja, S., Spakowitz, A. J., Zhang, Y., and Bustamante, C. (2006). Effect of force on mononucleosomal dynamics, *Proceedings of the National Academy of Sciences* **103**, pp. 15871–15876.

Morse, P. M. (1929). Diatomic molecules according to the wave mechanics. II. vibrational levels, *Physical Review* **34**, pp. 57–64.

Murphy, M. C., Rasnik, I., Cheng, W., Lohman, T. M., and Ha, T. (2004). Probing single-stranded DNA conformational flexibility using fluorescence spectroscopy, *Biophysical Journal* **86**, pp. 2530–2537.

Nakabachi, A., Yamashita, A., Toh, H., Ishikawa, H., Dunbar, H. E., Moran, N. A., and Hattori, M. (2006). The 160-kilobase genome of the bacterial endosymbiont carsonella, *Science* **314**, pp. 267–267.

Nasir, A., Forterre, P., Kim, K. M., and Caetano-Anollés, G. (2014). The distribution and impact of viral lineages in domains of life, *Frontiers in Microbiology* **5**, p. 194.

Nomidis, S. K., Kriegel, F., Vanderlinden, W., Lipfert, J., and Carlon, E. (2017). Twist-bend coupling and the torsional response of double-stranded DNA, *Phys. Rev. Lett.* **118**, p. 217801.

Perelroyzen, M. P., Lyamichev, V. I., Kalambet, A. Y., Lyubchenko, L. Y., and Vologodskii, A. D. (1981). A study of the reversibility of helix-coil transition in DNA, *Nucleic Acids Research* **9**, pp. 4043–4059.

Peterlin, A. (1953a). Light scattering by very stiff chain molecules, *Nature* **171**, pp. 259–260.

Peterlin, A. (1953b). Modèle statistique des grosses molécules à chaînes courtes. V. diffusion de la lumière, *Journal of Polymer Science* **10**, pp. 425–436.

Peyrard, M. and Bishop, A. R. (1989). Statistical mechanics of a nonlinear model for DNA denaturation, *Physical Review Letters* **62**, pp. 2755–2758.

Poland, D. and Scheraga, H. A. (1966). Occurrence of phase transition in nucleic acid models, *Journal of Chemical Physics* **45**, pp. 1464–1469.

Poland, D. and Scheraga, H. A. (1970). *Theory of helix-coil transitions in biopolymers* (Academic).

Printz, M. P. and von Hippel, P. H. (1965). Hydrogen exchange studies of DNA structure, *Proceedings of the National Academy of Sciences* **53**, pp. 363–370.

Purohit, P. K., Kondev, J., and Phillips, R. (2003). Mechanics of DNA packaging in viruses, *Proceedings of the National Academy of Sciences* **100**, pp. 3173–3178.

Richmond, T. J. and Davey, C. A. (2003). The structure of DNA in the nucleosome core, *Nature* **423**, pp. 145–150.

Rief, M., Clausen-Schaumann, H., and Gaub, H. E. (1999). Sequence-dependent mechanics of single DNA molecules, *Nature Structural Biology* **6**, pp. 346–349.

Rouzina, I. and Bloomfield, V. A. (2001). Force-induced melting of the DNA double helix 1. Thermodynamic analysis, *Biophysical Journal* **80**, pp. 882–893.

SantaLucia, J. (1998). A unified view of polymer, dumbbell, and oligonucleotide DNA nearest-neighbor thermodynamics, *Proceedings of the National Academy of Sciences* **95**, pp. 1460–1465.

Scalapino, D. J., Sears, M., and Ferrell, R. A. (1972). Statistical mechanics of one-dimensional Ginzburg-Landau fields, *Physical Review B* **6**, pp. 3409–3416.

Schildkraut, C., Marmur, J., and Doty, P. (1962). Determination of the base composition of deoxyribonucleic acid from its buoyant density in CsCl, *Journal of Molecular Biology* **4**, pp. 430–443.

Schindler, T., González, A., Boopathi, R., Roda, M. M., Romero-Santacreu, L., Wildes, A., Porcar, L., Martel, A., Theodorakopoulos, N., Cuesta-López, S., Angelov, D., Unruh, T., and Peyrard, M. (2018). Kinky DNA in solution: Small angle scattering study of a nucleosome positioning sequence, *Physical Review E* **98**, p. 042417.

Shimada, J. and Yamakawa, H. (1984). Ring-closure probabilities for twisted wormlike chains. application to DNA, *Macromolecules* **17**, pp. 689–698.

Silverstein, K. A. T., Haymet, A. D. J., and Dill, K. A. (2000). The strength of hydrogen bonds in liquid water and around nonpolar solutes, *Journal of the American Chemical Society* **122**, pp. 8037–8041.

Sim, A. Y. L., Lipfert, J., Herschlag, D., and Doniach, S. (2012). Salt dependence of the radius of gyration and flexibility of single-stranded DNA in solution probed by small-angle x-ray scattering, *Physical Review E* **86**, p. 021901.

Smith, D. E., Tans, S. J., Smith, S. B., Grimes, S., Anderson, D. L., and Bustamante, C. (2001). The bacteriophage ϕ 29 portal motor can package DNA against a large internal force, *Nature* **413**, pp. 748–752.

Smith, S. B., Cui, Y., and Bustamante, C. (1996). Overstretching B-DNA: the elastic response of individual double-stranded and single-stranded DNA molecules, *Science* **271**, pp. 795–799.

Smith, S. B., Finzi, L., and Bustamante, C. (1992). Direct mechanical measurements of the elasticity of single DNA molecules by using magnetic beads, *Science* **258**, pp. 1122–1126.

Spakowitz, A. J. and Wang, Z.-G. (2004). Exact results for a semiflexible polymer chain in an aligning field, *Macromolecules* **37**, pp. 5814–5823.

Stellwagen, N. C. (1981). Electric birefringence of restriction enzyme fragments of DNA: Optical factor and electric polarizability as a function of molecular weight, *Biopolymers* **20**, pp. 399–434.

Stepanow, S. and Schütz, G. M. (2002). The distribution function of a semiflexible polymer and random walks with constraints, *Europhysics Letters* **60**, pp.

546–551.

Tachibana, H., Ueno-Nishio, S., Gotoh, O., and Wada, A. (1982). Salt concentration dependence of thermal denaturation of restriction fragment DNAs from phiX174, *The Journal of Biochemistry* **92**, pp. 623–635.

Tao, Y., Olson, N. H., Xu, N. W., Anderson, D. L., Rossmann, M. G., and Baker, T. S. (1998). Assembly of a tailed bacterial virus and its genome release studied in three dimensions, *Cell* **95**, pp. 431–437.

Theodorakopoulos, N. (2006). Phase transitions in one dimension: are they *all* driven by domain walls? *Physica D* **216**, pp. 185–190.

Theodorakopoulos, N. (2008). DNA denaturation bubbles at criticality, *Physical Review E* **77**, p. 031919.

Theodorakopoulos, N. (2010). Melting of genomic DNA: predictive modeling by nonlinear lattice dynamics, *Physical Review E* **82**, p. 021905.

Theodorakopoulos, N. (2011). Bubbles, clusters and denaturation in genomic DNA: modeling, parametrization, efficient computation, *Journal of Nonlinear Mathematical Physics* **18**, pp. 429–447.

Theodorakopoulos, N. (2019). Thermodynamics of force-induced B-DNA melting: Single-strand discreteness matters, *Phys. Rev. E* **99**, p. 032404.

Theodorakopoulos, N., Dauxois, T., and Peyrard, M. (2000). Order of the phase transition in models of DNA thermal denaturation, *Physical Review Letters* **85**, pp. 6–9.

Theodorakopoulos, N. and Peyrard, M. (2012). Base pair openings and temperature dependence of DNA flexibility, *Physical Review Letters* **108**, p. 078104.

Theodorakopoulos, N., Peyrard, M., and MacKay, R. S. (2004). Nonlinear structures and thermodynamic instabilities in a one-dimensional lattice system, *Physical Review Letters* **93**, p. 258101.

Thomas, R. (1954). Recherches sur la dénaturation des acides desoxyribonucléiques, *Biochimica et Biophysica Acta* **14**, pp. 231–240.

Tyagi, S. and Kramer, F. R. (1996). Molecular beacons: Probes that fluoresce upon hybridization, *Nature Biotechnology* **14**, pp. 303–308.

Urabe, H. and Tominaga, Y. (1981). Low-frequency Raman spectra of DNA, *Journal of the Physical Society of Japan* **50**, pp. 3543–3544.

Vafabakhsh, R. and Ha, T. (2012). Extreme bendability of DNA less than 100 base pairs long revealed by single-molecule cyclization, *Science* **337**, pp. 1097–1101.

van Eijck, L., Merzel, F., Rols, S., Ollivier, J., Forsyth, V. T., and Johnson, M. R. (2011). Direct determination of the base-pair force constant of DNA from the acoustic phonon dispersion of the double helix, *Physical Review Letters* **107**, p. 088102.

van Mameren, J., Gross, P., Farge, G., Hooijman, P., Modesti, M., Falkenberg, M., Wuite, G. J. L., and Peterman, E. J. G. (2009). Unraveling the structure of DNA during overstretching by using multicolor, single-molecule fluorescence imaging, *Proceedings of the National Academy of Sciences* **106**, pp. 18231–18236.

von Hippel, P. H. and Wong, K.-Y. (1971). Dynamic aspects of native DNA structure: Kinetics of the formaldehyde reaction with calf thymus DNA,

Journal of Molecular Biology **61**, pp. 587–613.

von Meyenn, K. (1970). Rotation von zweiatomigen Dipolmolekülen in starken elektrischen Feldern, *Zeitschrift für Physik* **231**, pp. 154–160.

Wada, A., Tachibana, H., Gotoh, O., and Takanami, M. (1976). Long range homogeneity of physical stability in double-stranded DNA, *Nature* **263**, pp. 439–440.

Wartell, R. M. and Benight, A. S. (1985). Thermal denaturation of DNA molecules: a comparison of theory with experiments, *Physics Reports* **126**, pp. 67–107.

Weidlich, T., Lindsay, S., Rui, Q., Rupprecht, A., Peticolas, W., and Thomas, G. (1990). A Raman-study of low-frequency intrahelical modes in A-DNA,B-DNA, and C-DNA, *Journal of Biomolecular Structure and Dynamics* **8**, pp. 139–171.

Wilson, R. (1978). The dichroic tensor of flexible helices in a magnetic field, *Biopolymers* **17**, pp. 1811–1814.

Yan, J., Kawamura, R., and Marko, J. F. (2005). Statistics of loop formation along double helix DNAs, *Phys. Rev. E* **71**, p. 061905.

Yuan, C., Chen, H., Lou, X. W., and Archer, L. A. (2008). DNA bending stiffness on small length scales, *Physical Review Letters* **100**, p. 018102.

Zimm, B. H. (1948). Apparatus and methods for measurement and interpretation of the angular variation of light scattering; preliminary results on polystyrene solutions, *The Journal of Chemical Physics* **16**, pp. 1099–1116.

Zimm, B. H. and Bragg, J. R. (1959). Theory of the phase transition between helix and random coil in polypeptide chains, *Journal of Chemical Physics* **31**, pp. 526–535.

Index

All-or-nothing (AON) model, 79

bacteriophage, $\phi 29$, 29
bend, permanent, 144, 146
bending stiffness, 2, 26, 129
Benoit–Doty relationship, 6
Bernoulli numbers, 11
Bessel function, 9
Bessel function, modified spherical, 3, 8, 26
biomolecular beacons, 149
birefringence, magnetic, 132
bubble, aggregation, 135
bubbles, 89
Bubbles, PB model, 106

clusters, 89
contour length, 1
cyclization, 15

diatomic molecule, rigid, 23
digamma function, 105
Dirac delta function, 7
DNA overstretching, torsionally constrained, 72
DNA, base pair, lifetime, 94
DNA, base pairs, breathing of, 93
DNA, double-stranded, persistence length, 133
DNA, flexibility, temperature dependent, 129

DNA, force-extension curve, 68
DNA, force-extension relationship, 26
DNA, hairpins, 149
DNA, hairpins, excluded volume, 154
DNA, hairpins, thermodynamics, 151
DNA, light scattering from, 49
DNA, melting profiles, Carsonella ruddii, 126
DNA, melting profiles, pBR322, 123
DNA, melting profiles, T-7, 122
DNA, melting, force-induced, 70
DNA, melting, homogeneous, 62
DNA, melting, long chains, 61
DNA, melting, multistep, 62
DNA, melting, oligonucleotides, 55
DNA, neutron scattering from, 50
DNA, overstretching, 68
DNA, packaging, 29
DNA, scattering from, 39
DNA, thermal unbinding of, 53
DNA, unzipping, 65
DNA, unzipping, mean-field theory of, 66
DNA, unzipping, PB model, 108
DNA, unzipping, PBD model, 126
DNA, vibrations, low-frequency, 94
DNA, virus, 29
domain wall, 161, 164
domain wall, thermodynamic stability of, 164

elastic stretch modulus, ds-DNA, 69
elastic stretch modulus, ss-DNA, 69
elasticity, entropic, 19
end-to-end distance, 7
end-to-end distance,definition, 1
enthalpies, nearest-neighbor model, 58
enthalpy, Van't Hoff, 56
enthalpy, calorimetric, 56
entropies, nearest-neighbor model, 58
excluded volume, 14, 154

FJC, structure factor, 41
Fluorescence resonance energy transfer (FRET), 150
force, external, 19
force-extension curve, FJC, 20
force-extension curve, KP, 25
force-extension curve, WLC, 21
forward scattering, 40, 146
Fourier transform, 7
freely jointed chain (FJC), 2

Gauss–Legendre quadratures, 167
Gaussian chain, 11
GC content, melting temperature dependence on, 54, 60
Guinier, plot, 41, 142, 145, 147

Heisenberg, ferromagnetic chain, 3
helix, growth, 78
helix, initiation, 60, 78
helix, termination, 60
helix-coil, equilibrium, 55
helix-coil, transition, 77
histone, 35
hydration energy, virus, 32
hydrogen bonds, 54, 59

imino proton, exchange, 94
Ising model, 81

kink, 144, 146
Kratky–Porod (KP) model, 2
Kratky–Porod chain, Monte Carlo simulation, 157

Kratky–Porod model, heterogeneous, 130
Kratky–Porod, heterogeneous chain, 139
Kratky–Porod, structure factor, 40
Kuhn length, 21

Landau theorem, 161
Landau theorem, exceptions, 166
Landau theorem, inapplicable, 162
Legendre polynomial, 8
Legendre polynomial, associated, 8
length, characteristic, correlation, 106
length, characteristic, transverse, 104
loop entropy, 16, 84
loop formation, 15, 154
loop formation, probability of, 13

Monte Carlo simulation, 12
Monte Carlo simulation, Kratky–Porod chain, 157
Monte Carlo simulation, WLC, 158
Morse potential, 95

nucleosome, bending energy, 37
nucleosome, unwrapping, 37
nucleosome, wrapping, 35

PB model, 106
PB model, bubbles, 106
PB model, melting temperature, 103
PB model, optical phonons, 96
PB model, phase transition, 103
PBD model, first-order transition, effective, 114
PBD model, heterogeneity, 117
PBD model, melting profiles, 125
PBD model, nonlinear stacking interaction, 109
PBD model, thermally activated barrier, 111
PBD model, thermodynamic properties, 111
PBD model, unzipping, 126
Persistence length, 4

persistence length, double-stranded DNA, 27, 133
persistence length, single-stranded DNA, 28
Peyrard–Bishop (PB) model, 95
Peyrard–Bishop–Dauxois (PBD) model, 109
phase transition, continuous, 87
phase transition, first order, 88
phase transition, in one dimensional system, 90, 161
phonon, soft mode, 94
Poland–Scheraga (PS) model, 82
Poland–Scheraga, phase transition, 84
poly(A) chains, 156
poly(A) loops, 152
poly(T) chains, 156
poly(T) loops, 152
polylogarithm function, 85
polypeptides, synthetic, 77

radius of gyration, 5, 15, 41, 142, 145, 146
random walk, 2
ring closure, 13, 15
ring closure probability, hairpin loop, 155
rotator, quantum, 23

salt concentration, melting temperature dependence on, 60

scattering, elastic, 39
Schrödinger equation, 23, 102
self-avoiding walk (SAW), 15, 154
small angle neutron scattering (SANS), 50
small-angle X-ray scattering (SAXS), 156
soliton, topological, 164
stacking interaction, 59
Stark effect, 23

TI equation, 168
Transfer integral (TI), 99
Transfer integral, gradient expansion, 101
twisting rigidity, 26

virus, 29
virus, bending energy, 31
virus, hydration energy, 31
virus, packaging force, 31
virus, pressure, 33

WLC, Monte Carlo simulation, 158
WLC, structure factor, 43
wormlike chain (WLC), 2

Zimm plot, 50
zipper model, 79
zipper, generalized model, 80

www.ingramcontent.com/pod-product-compliance
Lightning Source LLC
Chambersburg PA
CBHW050627190326
41458CB00008B/2176